SMART URBANISM

All over the world, an emerging form of smart urbanism is increasingly presented as a vital response to meeting the future challenges of the urban condition. The promise of smart urbanism – from smart cities to smart infrastructures, buildings and grids – is that digital technologies can provide solutions to fundamental problems, from environmental sustainability and the governance of increasingly complex cities to service delivery and social inclusion. Key voices in these wide-ranging debates and initiatives include technology companies, national governments, supranational agencies and civil society organisations. All of these, in different ways, claim crucial roles for digital technologies across urban life, including managing and controlling infrastructures, achieving greater effectiveness in service demand and provision, reducing carbon emissions, developing greater social interaction and community networks and rebooting economic growth – all of this whilst increasing political transparency and rolling out innovative systems for holding authorities to account. But can cities be simply rebooted through the integration of digital technologies with buildings, neighbourhoods, infrastructures and people? Is the smart urbanism offer about new possibilities for the future transformation of our cities, or is it just a sophisticated form of 'business-as-usual', a mere tweaking of existing arrangements?

This book evaluates the contemporary promise of smart urbanism and critically appraises its potential. It brings together leading international researchers on smart cities to critically address these discourses and practices across the world, from India to the USA and from South Africa to Europe. It examines what drives smart city initiatives and to what ends, evaluates what new capabilities are being created by whom and with what exclusions, explores how smart city initiatives are being differently developed and contested, considers how these processes vary within and across cities and investigates their social, political, economic and environmental consequences. *Smart Urbanism* is an essential resource for advanced-level students and early-career researchers across urban studies.

Simon Marvin is Professor and Director of the Urban Institute at Sheffield University.

Andrés Luque-Ayala is a Lecturer in the Department of Geography at Durham University.

Colin McFarlane is a Reader in the Department of Geography at Durham University.

SMART URBANISM

Utopian vision or false dawn?

Edited by Simon Marvin, Andrés Luque-Ayala and Colin McFarlane

LONDON AND NEW YORK

First published 2016
by Routledge
2 Park Square, Milton Park, Abingdon, Oxon OX14 4RN

and by Routledge
711 Third Avenue, New York, NY 10017

Routledge is an imprint of the Taylor & Francis Group, an informa business

© 2016 Simon Marvin, Andrés Luque-Ayala and Colin McFarlane

The right of the editors to be identified as the authors of the editorial material, and of the authors for their individual chapters, has been asserted in accordance with sections 77 and 78 of the Copyright, Designs and Patents Act 1988.

All rights reserved. No part of this book may be reprinted or reproduced or utilised in any form or by any electronic, mechanical, or other means, now known or hereafter invented, including photocopying and recording, or in any information storage or retrieval system, without permission in writing from the publishers.

Trademark notice: Product or corporate names may be trademarks or registered trademarks, and are used only for identification and explanation without intent to infringe.

British Library Cataloguing in Publication Data
A catalogue record for this book is available from the British Library

Library of Congress Cataloging in Publication Data
Smart urbanism : utopian vision or false dawn? / edited by Simon Marvin, Andrés Luque-Ayala and Colin McFarlane.
  Includes bibliographical references and index.
  1. Urbanization. 2. Intelligent sensors. 3. Information technology. 4. City planning–Technological innovations. I. Marvin, Simon, 1963– editor. II. Luque-Ayala, Andrés, editor. III. McFarlane, Colin, 1979– editor.
  HT166.S58778 2016
  307.76–dc23   2015024486

ISBN: 978-1-138-84422-3 (hbk)
ISBN: 978-1-138-84423-0 (pbk)
ISBN: 978-1-315-73055-4 (ebk)

Typeset in Bembo
by Out of House Publishing

Printed and bound in Great Britain by
TJ International Ltd, Padstow, Cornwall

CONTENTS

List of illustrations vii
About the authors viii
Acknowledgements xii
Chapter credits xiv

1 Introduction 1
  Andrés Luque-Ayala, Colin McFarlane and Simon Marvin

2 Smart cities and the politics of urban data 16
  Rob Kitchin, Tracey P. Lauriault and Gavin McArdle

3 IBM and the visual formation of smart cities 34
  Donald McNeill

4 The smart entrepreneurial city: Dholera and 100 other utopias in India 52
  Ayona Datta

5 Getting smart about smart cities in Cape Town: beyond the rhetoric 71
  Nancy Odendaal

6 Programming environments: environmentality and citizen sensing in the smart city 88
  Jennifer Gabrys

7 Smart city initiatives and the Foucauldian logics of governing
   through code   108
   Francisco R. Klauser and Ola Söderström

8 Geographies of smart urban power   125
   Gareth Powells, Harriet Bulkeley and Anthony McLean

9 Test bed as urban epistemology   145
   Nerea Calvillo, Orit Halpern, Jesse LeCavalier and Wolfgang Pietsch

10 Beyond the corporate smart city? Glimpses of other
   possibilities of smartness   168
   Robert G. Hollands

11 Conclusion   185
   Colin McFarlane, Simon Marvin and Andrés Luque-Ayala

Index   192

LIST OF ILLUSTRATIONS

Figures

2.1	Dublin Dashboard	21
2.2	CASA London Dashboard	25
3.1	Rio de Janeiro's Center of Operations, COR	36
4.1	The actually existing Dholera region	58
4.2	The actually existing Dholera region	58
4.3	JAAG methods of social action	63
5.1	Smart city initiatives in Cape Town	83
6.1	Madrid: managing homes	95
6.2	Enabling technologies	99
6.3	Curitiba: citizen reporting	101
8.1	The low-voltage smart grid as imagined in the UK context	131
8.2	Solar SunFlowers	132
9.1	Songdo International Business District under construction	146
9.2	Seoul urban screens	150
9.3	Smart pole detail	152
9.4	Demonstration control room, City of Tomorrow, Songdo	152
9.5	Sensor data can occasionally reach conclusions that are self-evident	153
9.6	Songdo is constructed by dredging sand from the ocean flow	155
9.7	Model of Songdo in the City of Tomorrow showroom	156
9.8	Communal garden inside a park	158
9.9	Underground access in Songdo	162
9.10	Towers around 'Central Park'	163
10.1	An ecological community	178

Table

2.1	The apparatus and elements of a data assemblage	20

ABOUT THE AUTHORS

Editors

Simon Marvin is Professor and Director of the Urban Institute at Sheffield University, UK. His research interests focus on the changing relationships between cities and infrastructure networks. More recently his work has focused on comparative analysis of low carbon transitions and the interactions between smart technologies and urban contexts.

Andrés Luque-Ayala is a Lecturer in the Department of Geography at Durham University, UK. His research focuses on the socio-political dimensions of emerging infrastructural narratives and configurations in cities in the global South. As part of this, his work critically examines the emergence of a local governance of energy and the interface between digital technologies, ecological security and development modes.

Colin McFarlane is an urban geographer based at Durham University, UK. His work focuses on urban learning, informality and infrastructure in cities. This has included research on the politics of urban knowledge, urban sanitation and everyday life in informal neighbourhoods, especially in Mumbai, Cape Town and Kampala.

Contributors

Harriet Bulkeley is a Professor of Human Geography at Durham University, UK. Her research is concerned with environmental governance and focuses on theorising and explaining the processes and practices of governing the environment, the urban politics of climate change and sustainability and the political geographies of environmental governance.

Nerea Calvillo is an architect, lecturer, researcher and curator currently working on environmental monitoring and ontologies of the air. She is Design Critic in Architecture at the GSD Harvard University, USA, curator of the Connecting Cities Network European project on Media Facades and researcher at the Centre for Interdisciplinary Methodologies at Warwick University, UK.

Ayona Datta is a Senior Lecturer and Research Cluster Leader in Citizenship and Belonging at the School of Geography, University of Leeds, UK. Her research and writing broadly focuses on the gendered processes of citizenship and belonging and the politics of urbanisation across the global North and South. Ayona is author of *The Illegal City: Space, Law and Gender in a Delhi Squatter Settlement* (Ashgate, 2012). Her forthcoming co-edited books *Fast Cities: Mega-urbanization in the Global South* (Routledge) and *Ecological Citizenships in the Global South* (Zed Books) will be published in 2016.

Jennifer Gabrys is a Reader in the Department of Sociology at Goldsmiths, University of London, UK, and principal investigator on the ERC-funded project 'Citizen Sensing and Environmental Practice: Assessing Participatory Engagements with Environments through Sensor Technologies'. Her study on environmental sensing, *Program Earth: Environmental Sensing Technology and the Making of a Computational Planet*, is forthcoming from the University of Minnesota Press. Gabrys's work can be found at citizensense.net and jennifergabrys.net.

Orit Halpern is an Assistant Professor at the New School, New York City, USA. Her research is on histories of digital media, cybernetics, cognitive and neuroscience, art and design. She is the author of the book *Beautiful Data: A History of Vision and Reason since 1945* (Duke Press, 2014) and has published and created works for a variety of venues, including the *Journal of Visual Culture*, *Public Culture* and at ZKM in Karlsruhe, Germany.

Robert G. Hollands is a Professor of Sociology in the School of Geography, Politics and Sociology at Newcastle University, UK. He is the recent recipient of a Major Research Fellowship from the Leverhulme Trust entitled 'Urban Cultural Movements and the Struggle for Alternative Creative Spaces'.

Rob Kitchin is Professor and ERC Advanced Investigator at the National University of Ireland Maynooth. He is principal investigator of the Programmable City project, the Dublin Dashboard, the All-Island Research Observatory and the Digital Repository of Ireland, and the author of *Code/Space: Software and Everyday Life* (MIT Press, 2011) and *The Data Revolution* (Sage, 2014).

Francisco R. Klauser is Professor in Political Geography at the University of Neuchâtel, Switzerland. His work bridges the academic fields of human geography, surveillance studies and risk research. In recent years, he has developed an international portfolio of work dealing with the topics of sport mega-event security, risk and surveillance in the aviation sector, video surveillance and Big Data.

Tracey P. Lauriault is Assistant Professor of Critical Media Studies and Big Data at Carleton University, Canada, in the School of Journalism and Communication, a Research Associate of the ERC-funded Programmable City Project at the National University of Ireland Maynooth and a Research Associate of the Geomatics and Cartographic Research Centre in Ottawa. She is a member of the Government of Ireland Public Bodies Working Group on Open Data, the Dublin Region Homelessness Executive Research and Data Advisory Committee (RDAC) and the Research Data Alliance (RDA) Legal Interoperability Working Group.

Jesse LeCavalier is an architect with interests in logistics and urbanism. He is Assistant Professor in the College of Architecture and Design, New Jersey Institute of Technology, USA, and a member of Co+LeCavalier, a design studio concerned with transforming everyday life at a range of scales, including furniture, buildings and urban design. Most recently, he was a senior researcher at the Future Cities Laboratory as part of the Singapore-ETH Center for Global Environmental Sustainability.

Gavin McArdle is a Research Fellow at the National Centre for Geocomputation at the National University of Ireland Maynooth and an Industry Research Fellow at the IBM Smarter Cities Technology Centre in Dublin. His research interests are in the area of geographic information science and include urban dynamics and geovisual analysis.

Anthony McLean is based at the Department of Town and Regional Planning at Sheffield University, UK. He is concerned with the evolving social, political and economic contexts of urban infrastructures, how they are managed and maintained and how they are interacted with on a daily basis.

Donald McNeill is Professor of Urban and Cultural Geography at the Institute for Culture and Society, Western Sydney University, Australia. He is an Australian Research Council Future Fellow (2012–16) with a project entitled 'Governing Digital Cities', which develops a critical political economy of smart cities, start-up ecosystems and digital urban economic strategies.

Nancy Odendaal teaches urban planning at the University of Cape Town, South Africa. Her past research is on the role of technology in urban change and planning education in Africa. Her work now focuses on infrastructure and spatial change in cities of the global South.

Wolfgang Pietsch is a physicist-turned-philosopher of science at the Munich Center for Technology in Society, Technische Universität München, Germany. His research focuses on scientific method, especially fundamental concepts like causality and probability. Lately, he has become fascinated by novel methodological paradigms emerging in data-intensive science.

Gareth Powells is a Lecturer in Human Geography at Newcastle University, UK. His interdisciplinary research and teaching focuses on environmental and economic sustainability and how these issues connect with everyday life and urban geographies.

Ola Söderström is Professor of Social and Cultural Geography at the University of Neuchâtel, Switzerland. His work investigates the production and effects of urban material cultures with special interests in visual studies, policy mobility studies, science and technology studies, geographies of architecture and, more recently, geographies of mental health.

ACKNOWLEDGEMENTS

We would like to acknowledge the help and support of a number of people and organisations that have helped with the development of the ideas, networks and practical processes associated with the production of this book.

The book's origins can be traced to the international workshop *Smart Urbanism: Utopian Vision or False Dawn?*, funded by the Urban Studies Foundation and co-organised jointly by the three editors. The workshop, held at Durham University in June 2013, was one of the first international gatherings of academics and practitioners aimed at critically examining emerging practices and concepts of smart urbanism. It involved the participation of over twenty scholars from universities in Europe, the United Kingdom, South Africa, Brazil, Australia and the United States and four practitioners working in the public and private sectors delivering smart initiatives in different urban contexts. A full workshop report is available from www.urbanstudiesfoundation.org. For reasons of space, and to ensure the coherence of the volume, it was not possible to include all the papers from the workshop in the book. However, we would like to thank all the delegates for producing excellent papers and stimulating critical and constructive discussion on the topic of smart urbanism. We'd also like to thank the Urban Studies Foundation for their generous sponsorship of the workshop and the subsequent publication of a research agenda flowing from the workshop in the journal *Urban Studies*, and hope they are pleased with the book that their funding has helped produce.

Both the Geography Department and the Durham Energy Institute at Durham University supported the early development of ideas leading to the book, by funding a small pilot grant on smart urbanism. This involved mapping the dynamics of smart cities initiatives worldwide, alongside a literature review and web search aimed at identifying potential contributors to the workshop. This helped us produce a much better workshop specification and assemble the right set of contributors.

During the writing of the book we undertook site visits to two key smart city initiatives that helped shape our thinking about the phenomenon of smart urbanism. During 2012 we visited the IBM Smart Cities Laboratory in Dublin with colleagues from Durham University, and in 2015 we visited the UK Technology Strategy Board Future Cities Demonstrator in Glasgow with colleagues from Brazil and the UK. We'd like to acknowledge the support and cooperation of IBM Smart Cities Lab, Glasgow City Council, Strathclyde University and Urban Tide in facilitating these fascinating and revealing visits.

We are grateful to a wider set of colleagues at Durham University who have participated in numerous discussions about smart urbanism that have equally enthused and puzzled us! Why is it so difficult to understand this phenomenon? In particular we would like to thank Ben Anderson, Harriet Bulkeley, Mike Crang, Rachel Gordon and Ruth Machen. Thank you for helping to create an enquiring and stimulating environment for this project.

Finally, our thanks are due to Andrew Mould, our publisher at Routledge, for continued support on urban book projects, and to Sarah Gilkes, his editorial assistant, for her support throughout the writing process and in getting us to production.

CHAPTER CREDITS

Chapter 1   Introduction

The opening sections of this chapter draw on re-edited sections of the article Luque-Ayala, A. and Marvin, S. (2015) 'Developing a critical understanding of smart urbanism?' *Urban Studies* 52 (12): 2105–2116. The authors would like to thank Sage and the Editors of *Urban Studies* for permission to reproduce sections from this longer article here.

Chapter 2   Smart cities and the politics of urban data

This chapter draws heavily on two previously published papers: Kitchin, R. (2014) 'The real-time city? Big data and smart urbanism.' *GeoJournal* 79 (1): 1–14; and Kitchin, R., Lauriault, T. P. and McArdle, G. (2015) 'Knowing and governing cities through urban indicators, city benchmarking and real-time dashboards.' *Regional Studies, Regional Science* 2: 1–28.

Chapter 4   The smart entrepreneurial city: Dholera and 100 other utopias in India

This chapter draws heavily on a published paper: Datta, A. (2015) 'New urban utopias of postcolonial India: entrepreneurial urbanization in Dholera smart city, Gujarat.' *Dialogues in Human Geography* 5 (1): 3–22.

Chapter 6   Programming environments: environmentality and citizen sensing in the smart city

This is a condensed version of Gabrys, J. (forthcoming) 'Citizen sensing in the smart and sustainable city: from environments to environmentality' and is reprinted here by permission of the University of Minnesota Press from their forthcoming volume titled *Program Earth: Environmental Sensing Technology and the Making of a Computational Planet* (Copyright by the Regents of the University of Minnesota). In addition, an extended version of this chapter was originally published in Gabrys, J. (2014) 'Programming environments: environmentality and citizen sensing in the smart city.' *Environment and Planning D: Society and Space* 32 (1): 30–48.

Chapter 7   Smart city initiatives and the Foucauldian logics of governing through code

This is a re-edited version of a longer article: Klauser, F., Paasche, T. and Söderström, O. (2014) 'Michel Foucault and the smart city: power dynamics inherent in contemporary governing through code.' *Environment and Planning D: Society and Space* 32 (5): 869–85. The authors are very thankful to Pion Ltd (www.pion.co.uk) and to the editors of *Environment and Planning D: Society and Space* (www.societyandspace.com) for allowing the reproduction of the paper.

Chapter 9   Test bed as urban epistemology

Part of the content of this chapter was originally published in Halpern, O., LeCavalier, J., Calvillo, N. and Pietsch, N. (2013) 'Test-bed urbanism.' *Public Culture* 25 (2): 272–306.

Chapter 10   Beyond the corporate smart city? Glimpses of other possibilities of smartness

This chapter is a re-edited version of a longer article: Hollands, R. G. (2015) 'Critical interventions into the corporate smart city.' *Cambridge Journal of Regions, Economy and Society* 8 (1): 61–77. The authors would like to thank Oxford University Press for permission to reproduce sections from this longer article here.

1
INTRODUCTION

Andrés Luque-Ayala, Colin McFarlane and Simon Marvin

Smart urbanism is emerging at the intersection of visions for the future of urban places, new technologies and infrastructures. Promoted by international organisations, the corporate sector and national and local governments alike, the dominant vision is of the meshing of intelligent infrastructure, high-tech urban development, the digital economy and e-citizens. Discourses around smart urbanism are deeply rooted in seductive and normative visions of the future where technology stands as the primary driver for change. This novel form of urbanism, it is argued, provides a flexible and responsive means of addressing the challenges of urban growth and renewal, responding to climate change, increasing resilience, promoting sustainable economic growth and building a more socially inclusive society (European Commission, 2013b). Yet our understanding of the opportunities, challenges and implications of smart urbanism is limited. Research in this field is still at an early stage, fragmented along disciplinary lines (e.g. Hollands, 2008) and often based on single-city case studies (e.g. Mahizhnan, 1999; Bakıcı et al., 2013). As a result, we lack both the theoretical insight and empirical evidence required to assess the implications and potentially transformative consequences of how smart urbanisation is emerging in different urban contexts. This book critically addresses what new capabilities are being created, by whom and with what exclusions; how these are being developed – and contested; where this is happening, both within and between cities; and with what sorts of social and material consequences.

The aim of the book is to unpack the different logics and rationales behind smart urbanism discourses and proposals, and in this way to understand the ways by which imaginaries of the future are currently being constructed along with their socio-technical and political implications. The contributors explore the implications of the deployment of smart technologies and discourses in the city, their possible splintering or integrating natures, and their real potential for the delivery of the promise and the possibility of imagining alternative urban futures through the means unlocked by smart urbanisation.

Smart Urbanism: Utopian Vision or False Dawn? pursues its aims and ambition through three objectives. First, by developing an interdisciplinary conceptual approach for the analysis of emerging digital and smart forms of urbanisation. The book examines how smart urbanism is currently conceptualised within urban studies, identifying areas for agreement, dialogue and dissent. We consider what theorisations of the co-constitution of social and technical systems offer for the conceptualisation of the intersection between digital technologies, utopian computational narratives and the city.

Second, by generating new knowledge about the forms, dynamics and consequences of smart urbanism in an internationally comparative context. Existing work on smart urbanism is in its infancy, confined to particular disciplines and single cases. There is a lack of comparative analysis and a dearth of knowledge about the range of urban contexts within which forms of smart and digital urbanisation are emerging internationally. Such an extensive analysis of smart urbanism is required in order to analyse where, why, how, for whom and with what implications.

Finally, by analysing how specific urban conditions enable and constrain transitions towards smart urbanisation and support the co-production of alternative pathways. Understanding the potential and transformative implications of the transition to the smart and digital city, and the possibilities for creating alternative – more sustainable and socially inclusive – pathways requires the intensive examination of how smart urbanism is produced and reproduced in particular urban contexts. Far from being passive backdrops, cities variously complicate, enable, disrupt, resist and translate smart urbanism.

Rationale for the book

A new language of 'smartness' is reshaping debates about contemporary cities, along with a new set of programmes and practices that are intent on realising smart urbanism. This is visible in the importance given to 'smart cities' in the various strategic urban future pathways devised by the UK and the European Union (e.g. European Commission, 2013a; HM Government, 2013), the development of 'smart city initiatives' in Asia, Australia, the US and elsewhere (e.g. Government of India, 1999, 2014; United States Congress, 2009) and the emergence of dedicated teams aimed at developing business opportunities in smart urbanism projects within global engineering, telecommunications and utilities companies such as IBM, Cisco, Toshiba, Google, General Electric, Hitachi and others. Smart urbanisation is projected, often following normative or teleological approaches, as a solution brought to the present to deal with a series of urban maladies, such as issues of transport congestion, resource limitation, climate change and even the need to expand democratic access.

Taken together, these new drivers and programmes are creating a new lexicon through which the development of cities is being forged – smart cities, smart infrastructure, smart meters, smart buildings, smart districts and smart grids. While often radically different in ambition and scope, the shift from conventional to smart logics is accompanied by new expectations of network flexibility, demand responsiveness, green growth, new services and connected communities. These expectations, in turn, are driving investments and reshaping policy priorities leading to the accelerated roll-out of smart urbanisation globally.

Yet the potential, limitations and broader implications of this transformation have not been critically examined. Existing research in the field has focused on the technical, engineering and economic dimensions of smart systems (Al-Hader *et al.*, 2009; Paskaleva, 2011). This research tends to have a 'problem-solving' focus (Caragliu *et al.*, 2011), concerned with achieving optimal outcomes for smart systems under current technical, political and market conditions (Chourabi *et al.*, 2012; Komninos *et al.*, 2013) and with limited critical analysis (Hollands, 2008).

This lack of critical evaluation, compounded with an emphasis on technological solutions that disregard the social and political domains, is a vital omission. Evidence from the analysis of previous interventions in urban systems, including the development of grid-based infrastructures (Hughes, 1983; Nye, 1999), modernist urban planning (Sandercock and Lysiottis, 1998) and new urbanism (Harvey, 1997), suggests that cities play a critical role in the development of these technological transitions, unfolding an inherently complex and contested process – which often fails. At the same time, urban studies has tended to neglect the material, technological and environmental dimensions of cities (Monstadt, 2009), though there is growing interest in the political ecologies and cyborgian nature of cities (Gandy, 2005; Heynen *et al.*, 2006), the politics of urban infrastructures (Graham and Marvin, 2001; McFarlane and Rutherford, 2008), the dynamics of urban sustainability and low carbon transitions (Bulkeley *et al.*, 2014) and emerging forms of 'digital urbanism' (Crang and Graham, 2007). The ways in which the social, economic and political potential of smart urbanism is fundamentally produced with and through technologies remains beyond the reach of social science perspectives.

Within this context, a critical assessment of smart urbanism is needed. From one perspective, smart urbanisation may serve to further deepen the splintering of urban networks that dominated the last part of the twentieth century for many cities (Graham and Marvin, 2001), creating deep divides between those with access to 'smart' and those without. Alternatively, in some guises, smart urbanism may serve to promote more 'community', 'civic' or 'metropolitan' forms of urban life and resource provision (e.g. Map Kibera, n.d.; SENSEable City Lab, n.d.). Internationally comparative research is critical in order to develop a nuanced understanding of how and why this varies across urban contexts. Understanding these processes will enable us to consider the current trajectories of smart urbanism and to examine its potential in cities where it has yet to become established. The limits of current disciplinary approaches mean that addressing the critical challenges of smart urbanism cannot be achieved without a step-change in thinking that combines critical insight across disciplines and places.

The emerging landscape of critical research around smart urbanism

Whilst a systematic and critical examination of emerging forms of smart urbanism is long overdue, critical scholars within geography, computer science, architecture, urban studies and media studies have long unpacked the interface between computing, information communication technologies (ICT) and the city. Rich research agendas on the role of ICT in the production of urban space have been developed around the notions of cybercities (Boyer, 1992; Graham and Marvin, 1996; Graham, 1999, 2002, 2004), digital cities (Ishida, 2000), ubiquitous computing (Galloway, 2004, 2008) and urban informatics (Burrows, 2009; Foth, 2009; Foth et al., 2011). The primary concerns of these literatures are diverse, ranging from an evaluation of the role of code and software in shaping the city and its politics (Amin and Thrift, 2002; Thrift and French, 2002; Graham, 2005; Kitchin and Dodge, 2011) to an analysis of the impact of Wi-Fi and wireless technologies in shaping the urban experience (Forlano, 2009; Middleton and Bryne, 2011) and an examination of the broader implications of digitally enabled urban environments capable of 'sensing' (Crang and Graham, 2007; Shepard, 2009). The implications of ICT and urban media technologies for the promotion of democratic expressions in the city and the constitution of notions of the public are a common concern within this body of literature (Crang, 2010; de Waal, 2011; Powell, 2011), pointing to the key mediating role of digital technologies in reshaping our understanding of the city.

Engaging directly with narratives around smart and intelligent cities, an initial wave of social science research focused on their potential as systems of innovation and knowledge circulation (Komninos, 2002, 2008). Through the use of analytical models emphasising the development of partnerships between academia, business and government, smart cities have been seen – through a limited critical lens – as technological spaces where ICT operates as a key input towards regional innovation and economic development (Leydesdorff and Deakin, 2011; Caragliu et al., 2011). These highly normative and 'problem-solving' perspectives position the smart city as 'a strategic device to encompass modern urban production factors in a common framework, [… highlighting] the importance of Information and Communication Technologies' towards urban competitiveness and 'urban wealth' (Caragliu et al., 2011: 65). Over the past few years, a further practical and conceptual development of ideas around smart cities has added several layers of complexity to this already rich research landscape, which we examine here through three themes: the politics of smart cities; smart urban capacities and capabilities; and the emergence of novel ways of knowing the city.

The politics of smart cities

First, to counterbalance the early 'problem-solving' perspectives, scholars have called for a more critical analysis of the promises and potential of smart urbanism (Hollands, 2008; Luque et al., 2014), pointing to the need to acknowledge its roll-out as fundamentally a political endeavour, the relevance of exploring both its normative nature and the possibility of alternative models, and the value of advancing a comparative approach when unpacking smart urbanism across varied geographies (Luque-Ayala and Marvin, 2015). Departing from an understanding of technology as the *sine qua non* solution for urban problems, activists and academics alike are increasingly pointing to the possible role of smart urbanism in fuelling urban conflicts, furthering forms of urban splintering, promoting regressive forms of centralisation, empowering the empowered and advancing novel forms of surveillance (Townsend, 2013). The improbable aspirational visions of exemplar smart cities such as Songdo (South Korea), Masdar (UAE) or PlanIT Valley (Portugal) are being uncovered as a function of a limited understanding of the urban and its complexity (Greenfield, 2013). The smart city, as an imprecise, rhetorical and largely ideological label, is revealed as a new form of high-tech urban entrepreneurialism (Hollands, 2008). Emerging critical voices within this space are unpacking the power/knowledge dimensions of smart urbanism, the way in which this techno-utopian discursive construct promotes neoliberal rationalities and specific private interests and its functioning as a Foucauldian 'technology of government at a distance' (Vanolo, 2014). Smart city rationalities, often grounded within the corporate world (Söderström *et al.*, 2014; Hollands, 2015; see Chapter 10, this volume), are increasingly regulating and managing urban systems through software-mediated techniques, resulting in a transformation of the modes for governing everyday life as well as contemporary functionings of power, space and regulation (Klauser *et al.*, 2014; see Chapter 7, this volume).

Smart urban capacities and capabilities

A second domain where a critical understanding of smart urbanism is emerging goes beyond a discursive evaluation of the smart city, examining the specific techniques, technologies and material configurations involved in its make-up. Three socio-technical processes have received particular attention, all of which are reviewed in more detail below (as well as throughout this volume): the role of digital urban dashboards and big data, the use of distributed sensing technologies and the centralisation of urban functions via urban operations centres. Whilst big data advocates value the potential of data for real-time urban management and urban analytics, its critics point to its hidden politics, the intrinsic corporatisation of city governance involved in this model and the lock-ins and brittleness associated with new types of city services that depend on software and communication technologies (Kitchin, 2014; see Chapter 2, this volume). Critics also point to how the use of big data in the city presents significant challenges for issues of privacy and surveillance, further enhancing the development of the panoptic city (Klauser and Albrechtslund, 2014; Kitchin, 2014). Data, as a new urban utility, 'gradually becom[es] a part of how we see the world', changing notions of causality and understandings of space (Thrift, 2014: 1264). Similarly, urban digital dashboards imply novel and narrow ways of knowing the urban, where 'the city as visualised facts' puts forward an instrumental rationality 'open to manipulation by vested interests, [characterised by] often unacknowledged methodological and technical issues' (Kitchin *et al.*, 2015: 6).

Within smart urbanism, the use of distributed sensing technologies and the potential coupling of mobile technologies, urban infrastructures and environmental sensors has been championed as a possible exemplar strategy for the achievement of both environmental security and urban sustainability (for example, via air quality sensors embedded across networks of personal mobile devices) (see Arup, 2010; GSMA, 2014). In critically unpacking this form of digital sustainability, scholars have examined the implications for citizenship associated with this model. Gabrys, for example, highlights the risks of a smart city model that 'potentially delimit[s] urban "citizenship" to a series of actions focused on monitoring and managing data … recast[ing] who or what counts as a "citizen"' (Gabrys, 2014: 30; see Chapter 6, this volume). As with urban big data, the surveillance and security implications of ambient data collection – an active area of development for security and industry players – have been questioned, where the public becomes a passive supplier of environmental information potentially 'lead[ing] to discrimination against individuals or groups who are perceived as living in risky environments or possessing risky lifestyles' (Monahan and Mokos, 2013: 287).

Finally, the attention received by the centralisation and optimisation of urban functions via integrated urban operations centres – an iconic technique of the smart city, exemplified by IBM's Center of Operations (COR) in Rio de Janeiro – has unveiled the control room as a novel site for urban research with significant implications for how both infrastructures and the city as a whole are governed (Gordon *et al.*, 2014; Luque-Ayala and Marvin, forthcoming). Rio's COR provides a new vantage point for the city, combining several modes of visual perception achieved exclusively through software packages marketed to municipalities by ICT and electronics companies. Advancing such ways of operating in the city requires a form of simplification and standardisation (McNeill, this volume) that threatens urban specificity, making all cities seemingly the same. Known as Urban Operating Systems, these software packages not only develop novel and selective ways of seeing the city, but also sidestep its politics by reconfiguring problematic urban flows as a matter of operations and maintenance (Luque-Ayala and Marvin, forthcoming).

Ways of knowing the city

Finally, a critical feature of smart urbanism is the consolidation and expansion of relatively novel ways of knowing and thinking the city. In the interface between digital technologies and urbanism, the city comes to be known through data, algorithms, modelling and a combination of visual and media channels. Many of the technologies examined in the previous section also exemplify the emergence of new urban knowledges, imaginations and ways of perceiving, illustrated by, for example, the growing number of urban dashboards developed both by municipal authorities and other urban stakeholders (e.g. London, Dublin, Glasgow, Michigan,

Salt Lake City and Birmingham),[1] the rise of municipal and city-based open data platforms (e.g. Chicago, New York, Glasgow, Amsterdam, Sacramento, Philadelphia and Hong Kong)[2] and the establishment of municipal operations centres integrating a variety of urban functions, services and infrastructures (e.g. Rio de Janeiro, Glasgow).[3] Dashboards, data platforms and control rooms are constituted as new mediating points through which knowledge about the urban context is coordinated, produced and enacted, creating new control capacities and capabilities, many of which rest on the generation of distributed viewpoints affecting the power/knowledge equation of the city.

Critically, these ways of knowing and thinking the city do not originate within the recent explosion of ICT technologies in everyday life. Rather, they have longer histories, linked to earlier transformations in urban governance. On one hand, thinking the city through both data and modelling can be traced to the post-war period, particularly in the United States, when knowledge and techniques on computing and cybernetics – originally developed within military and defence industries – found a new market within urban planning (Light, 2003; see also Marvin and Luque-Ayala, 2014). On the other, knowing – and governing – the city through a combination of visual techniques dates back to the nineteenth century, when the arrival of illumination and electricity infrastructures to the city played a pivotal role in the rise of surveillance and the development of spectacle – and thus a new politics of discipline and capital; beyond that, these changing visibilities played a role in the promotion of a liberal subject operating under a form of governing based on freedom (Otter, 2008; see also McNeill, this volume). At the outset of the twenty-first century, smart urbanism adds new layers of complexity to these developments. The post-war efforts of urban planners to model, compute and integrate urban functions, already out of fashion by the late 1970s, have made a vigorous comeback – this time highly informed by corporate rationalities, as logistics software and business integration packages developed throughout the 1980s and 1990s are repackaged as Urban Operating Systems (Marvin and Luque-Ayala, 2014). Knowing and seeing the city, and transforming urban governance through such capabilities, is no longer about lighting and electrification, but about integrating hundreds of traffic and surveillance cameras, real-time mapping urban services via GIS-enabled infrastructures and crowdsourcing information about urban environments through distributed networks composed of millions of individually owned mobile phone devices.

Such ways of knowing and thinking the city are highly rational, affective and mediatic at the same time. Open data platforms and dashboards are grounded in rational epistemologies, creating a false illusion of neutrality whilst suggesting that they are devoid of politics (Kitchin, this volume). In practice, they operate largely through a problematisation of the future via pre-emptive and speculative modes of action (cf. Anderson, 2010; Amoore, 2013): open data platforms make large amounts of data available for the public, making it possible to carry out multiple and yet unimagined data recombinations, arguably providing innumerable possibilities for innovation in public services. Interviews with municipal officers involved in the development of open data platforms reveal how such initiatives are more often described in terms of possibilities and potential than in terms of actual achievements. The resulting innovations, to be achieved through highly publicised hackathons or other forms of voluntary engagement of 'civic technologists' or 'civic hackers' (such as New York's NYCBigApps and Rio de Janeiro's Hackathon 1746 competitions), are sometimes elusive; with the exception of large cities which host a well-developed IT community, there is a limited understanding of the real potential for urban innovation provided by municipal open data platforms. Taking the speculative nature of open data platforms one step further, the City of Chicago is building the first real-time open-source predictive analytics platform, an initiative 'which will fundamentally change the way cities use data to improve services' (City of Chicago, 2013: 1). Such engagement with an emerging world of urban possibilities is as rational as it is affective, as it appeals to how we aspire for a particular type of efficient and optimised urban future.

Affect, anticipation and mediality come together in other type of exemplar smart urbanism intervention: integrated operations centres, such as the already mentioned Rio de Janeiro Center of Operations (COR). Rio's COR is characterised by a constant presence in the public eye via traditional media channels (TV and radio) as well as social media (including Twitter, Facebook and Waze). Its use of aerial imagery, maps and video cameras creates an illusion of total control. Yet such sense of total control exists only in anticipation of urban breakdown: alerts against risky weather conditions leading to landslides or flooding, real-time adjustments to traffic flows to avoid an inevitable congestion and urban reconfigurations to bypass the disruptions caused by urban protest (Luque-Ayala and Marvin, forthcoming). Broadly illustrating how we come to know the city within smart urbanism, the mediated viewing of the urban generated by Rio's COR – where an inbuilt sense of anticipation plays out through media operations (cf. Grusin, 2010) – modulates both the public's attention and its affects: the everyday of the city is seen in a permanent state of emergency, urgency becomes the paradigm of action and the immediate moment the primary focus of intervention (Luque-Ayala and Marvin, forthcoming).

Structure of the book

Each of the book's core chapters addresses a distinct dimension of smart urbanism. Chapter 2, by Kitchin *et al.*, opens the discussion by succinctly introducing the five most common critiques of smart cities: the promotion of technocratic and corporatised forms of governance; the creation of buggy, brittle and hackable urban systems; the implementation of forms of panoptic surveillance, predictive profiling and social sorting; and a false portrayal of data and algorithms as objective and non-ideological. Elaborating on the latter critique, the chapter provides an in-depth analysis of urban data through an examination of urban indicators, city benchmarking and real-time dashboards. Kitchin *et al.* uncover the claims of a realist epistemology – 'that claims to show the city as it actually is' (p. 29) – evidencing an instrumental rationality charged with politics and technical issues. Chapter 3 continues unpacking the claims to objectivity, truth and evidence embedded within smart city narratives. Here, McNeill draws on IBM's Smarter Cities campaign to argue that visual technologies are central to the ontological and practical configuration of smart cities, where only cities that can be seen through different and varied modes of perception are smart – whilst cities that cannot be viewed cannot be made smart. One of the key contributions of the chapter is to broaden the historical underpinnings of the smart city beyond the narrow frame of a recent 'digital era', linking it to nineteenth-century transformations in city government where the state reconfigures its ability to govern through practices of infrastructural visualisation (Otter, 2008; Crary, 2001). McNeill's analysis evidences how the smart city establishes new modes of governing through introspection (the state sees itself), totalising viewpoints (via synopsis), supervision (monitoring and inspection) and foresight (envisioning the future).

Chapters 4 and 5 focus on the ways in which smart urbanism is being rolled out in cities of the global South, specifically in India and South Africa. In Chapter 4, Datta provides a critical analysis of India's 2014 announcement that it would build 100 smart cities. This new configuration of Indian urbanism overlaps with narratives around innovation and enterprise as well as claims to modernity and development, 'essentially measured by [its] contribution in pushing up India's GDP and widening India's footprint in the global economy' (p. 52). Datta focuses on the social and political consequences of the initiative. An analysis of Dholera – allegedly India's first smart city – reveals that India's smart urbanism is a contingent process operating through land accumulation by dispossession, a form of politics interweaving dispossession, modernisation and liberalisation. In the case of South Africa, the focus of Chapter 5, Odendaal examines the incorporation of e-governance and digital infrastructure development into urban objectives. In Cape Town, after an early wave of digital urbanity tied to both city marketing and more inclusionary forms of development (Odendaal, 2006), a resurgence of smart city discourses has been linked to interventions by large ICT companies such as IBM and Cisco. Yet municipal governments appear as the primary agents in promoting technology as a social enabler. In contrast to the depiction of the Indian smart city of the previous chapter, Odendaal argues that smart urbanism initiatives under the leadership of municipal governments can play an important role in expanding democratic access and fulfilling a commitment to social development. In the context of unmet basic needs, low Internet access and a constitutional mandate for social empowerment, the smart city from the bottom-up can play a role in advancing the demands of social movements, supporting local government accountability and enabling social mobilisation.

Chapters 6 and 7, drawing on Foucauldian approaches, focus on the power/knowledge and governmentality implications of the smart city. In Chapter 6, Gabrys examines the use of sensor-based ubiquitous computing across urban infrastructures and mobile devices towards increasing urban sustainability. Through a case study of MIT and Cisco's Connected Sustainable Cities project, Gabrys argues that smart urbanism can potentially reconfigure (and constrain) our understanding of citizenship, where both cities and citizens become functional datasets to be managed and manipulated. Using Foucault's notion of environmentality – the distribution of governance within and through environments and environmental technologies – the chapter evaluates how understandings and practices of citizenship emerge within the smart city. Citizens, rather than governable subjects or populations, become an active technology of power playing a governing role via the provision of environmental data. Chapter 7, by Klauser and Söderström, focuses on the implications of governing the city through code/software, particularly via the management-at-a-distance of urban infrastructures via information technologies. Through an analysis of two case studies of smart energy management in Switzerland, the chapter asks theoretical questions around the role of IT in the governing of everyday life and the power dynamics characteristic of the smart city. These are answered through a framework that builds on Foucault's distinction between apparatuses of discipline and security (Foucault, 2009), examining three levels of interaction: referentiality (how does smart power relate to uncertainty, an inherent characteristic of the governing of multiplicities?); normativity and regulation (how are norms established and an idea of 'normality' conceived within smart initiatives?); and spatiality (how does the spatial organisation of the smart city mediate the exercise of power?). Klauser and Söderström suggest that Foucault's notion of apparatuses of security is useful for uncovering the aims, rationalities and transformations of power within smart urbanism. To paraphrase the authors, this smart energy case study shows how digital urban infrastructures are involved in constant processes of optimisation, where the balances between consumption and production are always being adjusted within acceptable limits, and where 'the aims and conditions of governing are constantly readapted and redefined' (p. 121).

In Chapter 8, Powells *et al.* take this analysis of smart energy models one step further, through an examination of how the very materiality of the electricity grid – power cables, gas pipes, buildings, others – is recombined and re-purposed around new rationales embedded in the digital and non-digital materialities of smart urbanism – such as smart meters, data, data loggers, control rooms and dashboards. This empirical work on energy networks, based on case studies from the UK and US, shows how the making of the smart grid is an unequal process. Specific political, social, economic and environmental priorities are foregrounded whilst others are left in the shadows. The resulting 'uneven power geometries' will mean that the experience of the smart grid will not be equal for all citizens. In its creation of energy consumers that are also producers (also known as 'prosumers'), and through a set of neoliberal apparatuses and techniques, the smart grid extends market rationalities and techniques into everyday life. Both transforming and consolidating energy markets, the smart grid demands a novel understanding of the role and responsibilities of the state (and the broader 'urban assemblage') in the provision and regulation of essential services.

The last two empirical chapters speak about the future of smart urbanism from contrasting apocalyptic and hopeful perspectives. In Chapter 9, Calvillo *et al.* discuss Songdo, a newly built 'smart city' in South Korea, where Cisco has played a key role through the provision of digital connectivity and ubiquitous computing infrastructures. Songdo is portrayed as an example of new forms of urban digital experimentation: speculative, smart and sentient, where all urban forms and beings are to be digitally interconnected. In this new urban world, data drives urban transformations, hierarchies and a rearrangement of urban life. Yet as an urban future that already feels obsolete, the half-built Songdo represents new urban ontologies that are digital and algorithmic, abstract (i.e. based on data densities, clouds, statistical risks or visibilities) and oppressively concrete (i.e. via ubiquitous cameras, secretive control rooms and windowless data centres). In Chapter 10, offering a contrasting account, Hollands explore the possibility of smart urbanism beyond corporate imaginations. Starting with a critique of the collapse of 'smart', competitiveness and urban entrepreneurialism, the chapter is a final reminder of the ideological nature of the idea of the smart city – neither technology nor its corporate urban reincarnation will automatically make cities more prosperous, efficiently governed, less environmentally wasteful or equal. However, in opening up alternative understandings of the smart city, the chapter offers an account of 'more modest and small-scale socio-technological interventions' (p. 176), where technology in itself is not a priori progressive but collaboratively rolled out in democratic ways in support of progressive ideas. Ideas which might support our efforts to imagine different ways 'of thinking about and "doing" smartness' (p. 176).

The final chapter, Chapter 11, presents the conclusions, identifying the key implications of the book for urban theory, urban governance and the methodological challenges of researching smart urbanism. Here, the book reflects on the opening of the 'black box' of smart urbanism, in order to discuss what is really going on here and what it might amount to for urban politics, economy, environment and everyday life.

Notes

1. See, respectively, London's City Dashboard (citydashboard.org/london/); the Dublin Dashboard portal (www.dublindashboard.ie/); the tourism dashboard developed by the independent Glasgow Economic Commission (http://glasgowtourismstrategy.com/dashboard/destination/); Michigan's Mi Dashboard (midashboard.michigan.gov); Salt Lake City's Sustainable City Dashboard (http://dotnet.slcgov.com/PublicServices/Sustainability/); and Birmingham's Civic Dashboard (http://civicdashboard.org.uk/).
2. See, respectively, the City of Chicago Data Portal (data.cityofchicago.org); New York City's NYC Open Data (nycopendata.socrata.com); Glasgow Open Data (data.glasgow.gov.uk); Amsterdam Open Data (www.amsterdamopendata.nl); the City of Sacramento Open Data Platform (data.cityofsacramento.org); OpenDataPhilly (www.opendataphilly.org); and Open Data Hong Kong (opendatahk.com/).
3. See Rio de Janeiro's Centro de Operações (www.centrodeoperacoes.rio.gov.br) and Glasgow's Integrated Operations Centre (futurecity.glasgow.gov.uk/index.aspx?articleid=10252).

References

Al-Hader, M., Rodzi, A., Sharif, A. R. and Ahmad, N. (2009) Smart city components architecture. In *International Conference on Computational Intelligence, Modelling and Simulation, 2009. CSSim'09*. Brno: IEEE, pp. 93–97. Available at: http://ieeexplore.ieee.org/xpls/abs_all.jsp?arnumber=5350055&tag=1 [Accessed July 2015].

Amin, A. and Thrift, N. (2002) *Cities: Reimagining the Urban*. Cambridge: Polity Press.
Amoore, L. (2013) *The Politics of Possibility*. Durham, NC: Duke University Press.
Anderson, B. (2010) Preemption, precaution, preparedness: anticipatory action and future geographies. *Progress in Human Geography* 34 (6): 777–798.
Arup (2010) Smart cities: transforming the 21st century city via the creative use of technology [online]. London. Available at: http://publications.arup.com/Publications/S/Smart_Cities.aspx [Accessed 25 May 2015].
Bakıcı, T., Almirall, E. and Wareham, J. (2013) A smart city initiative: the case of Barcelona. *Journal of the Knowledge Economy* 4 (2): 135–148.
Boyer, M. C. (1992) The imaginary real world of cybercities. *Assemblage* (18): 115–127.
Bulkeley, H., Castan-Broto, V. and Maassen, A. (2014) Low-carbon transitions and the reconfiguration of urban infrastructure. *Urban Studies* 51 (7): 1471–1486.
Burrows, R. (2009) Afterword: urban informatics and social ontology. In M. Foth (ed.) *Handbook of Research on Urban Informatics: The Practice and Promise of the Real-Time City*. Hershey, PA: Information Science Reference, pp. 450–453.
Caragliu, A., Del Bo, C. and Nijkamp, P. (2011) Smart cities in Europe. *Journal of Urban Technology* 18 (2): 65–82.
Chourabi, H., Nam, T., Walker, S., Gil-Garcia, J. R., Mellouli, S., Nahon, K., Pardo, T. A. and Scholl, H. J. (2012) Understanding smart cities: an integrative framework. In *System Science (HICSS), 2012 45th Hawaii International Conference*. Maui, HI: IEEE, pp. 2289–2297. Available at: http://ieeexplore.ieee.org/xpls/abs_all.jsp?arnumber=6149291 [Accessed July 2015].
City of Chicago (2013) Chicago named as one of five winners in Bloomberg Philanthropies' Mayors Challenge [online]. Available at: www.cityofchicago.org/city/en/depts/mayor/press_room/press_releases/2013/march_2013/chicago_named_asoneoffivewinnersin-bloombergphilanthropiesmayorsc.html [Accessed 25 May 2015].
Crang, M. (2010) Cyberspace as the new public domain. In C. W. Kihato, M. Massoumi, B. A. Ruble, P. Subirós and A. M. Garland (eds) *Urban Diversity: Space, Culture and Inclusive Pluralism in Cities Worldwide*. Baltimore, MD: Johns Hopkins University Press, pp. 99–122.
Crang, M. and Graham, S. (2007) Sentient cities: ambient intelligence and the politics of urban space. *Information, Communication & Society* 10 (6): 789–817.
Crary, J. (2001) *Suspension of Perception: Attention, Spectacle, and Modern Culture*. Cambridge, MA: MIT Press.
de Waal, M. (2011) The ideas and ideals in urban media. In M. Foth *et al.* (eds) *From Social Butterfly to Engaged Citizen: Urban Informatics, Social Media, Ubiquitous Computing, and Mobile Technology to Support Citizen Engagement*. Cambridge, MA: MIT Press, pp. 5–20.
European Commission (2013a) Energy technologies and innovation: Communication from the Commission to the European Parliament, the Council, the European Economic and Social Committee and the Committee of the Regions. COM (2013) 253, Brussels.
European Commission (2013b) *European Innovation Partnership on Smart Cities and Communities: Strategic Implementation Plan*. Brussels: EC.
Forlano, L. (2009) WiFi geographies: when code meets place. *Information Society* 25 (5): 344–352.
Foth, M. (ed.) (2009) *Handbook of Research on Urban Informatics: The Practice and Promise of the Real-Time City*. Hershey, PA: Information Science Reference.
Foth, M., Choi, J. H. and Satchell, C. (2011) Urban informatics. In *Proceedings of the ACM 2011 Conference on Computer Supported Cooperative Work. Hangzhou, China*. New York: ACM, pp. 1–8.

Foucault, M. (2009) *Security, Territory, Population: Lectures at the Collège de France 1977–1978*. New York: Picador.

Gabrys, J. (2014) Programming environments: environmentality and citizen sensing in the smart city. *Environment and Planning D: Society and Space* 32 (1): 30–48.

Galloway, A. (2004) Intimations of everyday life: ubiquitous computing and the city. *Cultural Studies* 18 (2–3): 384–408.

Galloway, A. (2008) A brief history of the future of urban computing and locative media. Unpublished PhD thesis, Carleton University, Ottawa, Ontario.

Gandy, M. (2005) Cyborg urbanization: complexity and monstrosity in the contemporary city. *International Journal of Urban and Regional Research* 29 (1): 26–49.

Gordon, R., Anderson, B., Crang, M. and Marvin, S. (2014) Controlling networks: Modes of governing infrastructural assemblages [Working paper]. Durham: Durham University.

Government of Australia (2009) *Smart Grid, Smart City: A New Direction for a New Energy Era*. Canberra: Department of the Environment, Water, Heritage and the Arts.

Government of India (2014) Draft concept note on smart city scheme. Delhi: Ministry of Urban Development.

Graham, S. (1999) Towards urban cyberspace planning: grounding the global through urban telematics policy and planning. In J. Downey and J. McGuiga (eds) *Technocities*. London: Sage, pp. 9–33.

Graham, S. (2002) Bridging urban digital divides? Urban polarisation and information and communications technologies (ICTs). *Urban Studies* 39 (1): 33–56.

Graham, S. (ed.) (2004) *The Cybercities Reader*. London: Routledge.

Graham, S. (2005) Software-sorted geographies. *Progress in Human Geography* 29 (5): 562–580.

Graham, S. and Marvin, S. (1996) *Telecommunications and the City*. London: Routledge.

Graham, S. and Marvin, S. (2001) *Splintering Urbanism: Networked Infrastructures, Technological Mobilities and the Urban Condition*. London: Routledge.

Greenfield, A. (2013) *Against the Smart City*. New York: Do Projects.

Grusin, R. (2010) *Premediation: Affect and Mediality after 9/11*. New York: Palgrave Macmillan.

GSMA (2014) *Mobile Smart City Benchmarking Report: Summary of Mobile Smart City Best Practice for Partnerships between Operators, Vendors and Government*. London: Groupe Speciale Mobile Association.

Harvey, D. (1997) The new urbanism and the communitarian trap. *Harvard Design Magazine* (1): 68–69.

Heynen, N., Kaika, M. and Swyngedouw, E. (eds) (2006) *In the Nature of Cities: Urban Political Ecology and the Politics of Urban Metabolism*. London: Taylor & Francis.

HM Government (2013) *Information Economy Strategy. Industrial Strategy: Government and Industry in Partnership*. London: Department for Business, Innovation and Skills.

Hollands, R. (2008) Will the real smart city please stand up? Intelligent, progressive or entrepreneurial? *City* 12 (3): 303–320.

Hollands, R. (2015) Critical interventions into the corporate smart city. *Cambridge Journal of Regions, Economy and Society* 8 (1): 61–77.

Hughes, T. (1983) *Networks of Power: Electrification in Western Society, 1880–1930*. Baltimore, MD: Johns Hopkins University Press.

Ishida, T. (2000) Understanding digital cities. In T. Ishida and K. Isbister (eds) *Digital Cities*. New York: Springer, pp. 7–17.

Kitchin, R. (2014) The real-time city? Big data and smart urbanism. *GeoJournal* 79 (1): 1–14.

Kitchin, R. and Dodge, M. (2011) *Code/Space: Software and Everyday Life*. Cambridge, MA: MIT Press.

Kitchin, R., Lauriault, T. P. and McArdle, G. (2015) Knowing and governing cities through urban indicators, city benchmarking and real-time dashboards. *Regional Studies, Regional Science* 2 (1): 6–28.

Klauser, F. and Albrechtslund, A. (2014) From self-tracking to smart urban infrastructures: towards an interdisciplinary research agenda on Big Data. *Surveillance & Society* 12 (2): 273–286.

Klauser, F., Paasche, T. and Söderström, O. (2014) Michel Foucault and the smart city: power dynamics inherent in contemporary governing through code. *Environment and Planning D: Society and Space* 32 (5): 869–885.

Komninos, N. (2002) *Intelligent Cities: Innovation, Knowledge Systems, and Digital Spaces.* London: Taylor & Francis.

Komninos, N. (2008) *Intelligent Cities and Globalisation of Innovation Networks.* London: Routledge.

Komninos, N., Pallot, M. and Schaffers, H. (eds) (2013) Special issue: Smart cities and the future internet in Europe. *Journal of the Knowledge Economy* 4 (2): 119–134.

Leydesdorff, L. and Deakin, M. (2011) The triple-helix model of smart cities: a neo-evolutionary perspective. *Journal of Urban Technology* 18 (2): 53–63.

Light, J. (2003) *From Warfare to Welfare: Defense Intellectuals and Urban Problems in Cold War America.* Baltimore, MD: Johns Hopkins University Press.

Luque, A., McFarlane, C. and Marvin, S. (2014) Smart urbanism: cities, grids and alternatives. In M. Hodson and S. Marvin (eds) *After Sustainable Cities?* London: Routledge, pp. 74–90.

Luque-Ayala, A. and Marvin, S. (2015) Developing a critical understanding of smart urbanism? *Urban Studies* 52 (12): 2105–2116.

Luque-Ayala, A. and Marvin, S. (forthcoming) The maintenance of urban circulation: an operational logic of infrastructural control. *Environment and Planning D: Society and Space.*

Mahizhnan, A. (1999) Smart cities: the Singapore case. *Cities* 16 (1): 13–18.

Map Kibera (n.d.) Putting marginalized communities on the map [online]. Available at: http://mapkibera.org [Accessed 20 May 2014].

Marvin, S. and Luque-Ayala, A. (2014) Urban operating systems: diagramming the city [working paper]. Durham: Durham University.

McFarlane, C. and Rutherford, J. (2008) Political infrastructures: governing and experiencing the fabric of the city. *International Journal of Urban and Regional Research* 32 (2): 363–374.

Middleton, C. A. and Bryne, A. (2011) An exploration of user-generated wireless broadband infrastructures in digital cities. *Telematics and Informatics* 28 (3): 163–175.

Monahan, T. and Mokos, JT. (2013) Crowdsourcing urban surveillance: the development of homeland security markets for environmental sensor networks. *Geoforum* 49: 279–288.

Monstadt, J. (2009) Conceptualizing the political ecology of urban infrastructures: insights from technology and urban studies. *Environment and Planning A* 41: 1924–1942.

Nye, D. E. (1999) *Consuming Power: A Social History of American Energies.* Cambridge, MA: MIT Press.

Odendaal, N. (2006) Towards the digital city in South Africa: issues and constraints. *Journal of Urban Technology* 13 (3): 29–48.

Otter, C. (2008) *The Victorian Eye: A Political History of Light and Vision in Britain, 1800–1910.* Chicago: University of Chicago Press.

Paskaleva, K. A. (2011) The smart city: a nexus for open innovation? *Intelligent Buildings International* 3 (3): 153–171.

Powell, A. (2011) Metaphors, models and communicative spaces: designing local wireless infrastructure. *Canadian Journal of Communication* 36 (1): 91–114.

Sandercock, L. and Lysiottis, P. (1998) *Towards Cosmopolis: Planning for Multicultural Cities.* New York: J. Wiley & Sons.

SENSEable City Lab. (n.d.) Forage tracking [online]. Available at: http://senseable.mit.edu/foragetracking/ [Accessed 14 May 2014].

Shepard, M. (2009) Sentient city survival kit: archaeology of the near future. Presented at Digital Arts and Culture Conference, University of California, Irvine, 12–15 December 2009. Available at: http://escholarship.org/uc/item/4zp0c4x2 [Accessed 7 August 2015].

Söderström, O., Paasche, T. and Klauser, F. (2014) Smart cities as corporate storytelling. *City* 18 (3): 307–320.

Thrift, N. (2014) The promise of urban informatics: some speculations. *Environment and Planning A* 46 (6): 1263–1266.

Thrift, N. and French, S. (2002) The automatic production of space. *Transactions of the Institute of British Geographers* 27 (3): 309–335.

Townsend, A. (2013) *Smart Cities: Big Data, Civic Hackers, and the Quest for a New Utopia.* New York and London: W.W. Norton.

United States Congress (2009) *American Recovery and Reinvestment Act of 2009.* Washington, DC: Government Printing Office.

Vanolo, A. (2014) Smartmentality: the smart city as disciplinary strategy. *Urban Studies* 51 (5): 883–898.

2
SMART CITIES AND THE POLITICS OF URBAN DATA

Rob Kitchin, Tracey P. Lauriault and Gavin McArdle

Introduction

'Smart city' seems to be the urban buzzword for the 2010s. The ambition for just about every city across the planet appears to be to become 'smarter'. Indeed, some nations have actively embraced the notion, with India, for example, announcing that it is to build 100 smart cities over the coming decades to accommodate a rapidly growing urban population (see Chapter 4). However, what constitutes a smart city is not universally agreed upon. In general terms, there are two main visions of smart urbanism, both of which are underpinned by the roll-out of new information and communication technologies (ICTs) and neoliberal visions of market-led and technocratic solutions to city governance and development, and are promoted as pragmatic, non-ideological and commonsensical in approach (Kitchin, 2014a).

On the one hand, a smart city is one whose urban fabric is increasingly instrumented, composed of 'everyware' (Greenfield, 2006) – software-enabled infrastructures and networked digital devices and sensors that are used to augment urban management and governance. Here, a smart city is one that can be monitored, managed and regulated in real time using ICT infrastructure and ubiquitous computing that generate big data (Townsend, 2013). On the other hand, a smart city is one whose economy is increasingly driven by technology-inspired innovation and entrepreneurship that, in turn, will attract businesses and jobs, create efficiencies and savings and raise the productivity and competitiveness of government and businesses (Caragliu *et al.*, 2009). Here, the focus is on the formulation and adoption of policies that use ICT to reshape human capital, creativity, education, sustainability, governance and economic activity to produce knowledge-driven, competitive, resilient urban systems. In many cases, cities are pursuing becoming smart in both regulatory and economic terms.

Whilst the creation of smart cities has many supporters, most notably governments that hope to address and manage the many issues cities face using

ICT-based solutions and businesses that seek to profit from selling new smart city technologies and services, smart urbanism has not been universally welcomed. Indeed, a number of critical scholars and community activists have challenged the prevailing rhetoric and have sought to unpack, contextualise and make sense of smart city initiatives (Hollands, 2008; Greenfield, 2013; Vanolo, 2014; Datta, 2015), or develop a more inclusive notion of a smart city (Hill, 2013; Townsend, 2013). This chapter examines five critiques of smart cities in broad terms, followed by a more sustained discussion of one of these critiques, namely the politics of urban digital data and the development of urban indicators, city benchmarking and real-time dashboards and their use in urban governance. Such a focus on urban digital data is important because its generation, exchange and analysis is central to the production of smart urbanism: the material that networked ICT systems process and from which they create and extract value. The key thrust of the argument developed is that whilst the smart city technologies and initiatives are generally portrayed and positioned as technical, pragmatic, commonsensical and non-ideological – that is, as rational interventions designed to improve social, economic and governance systems – they are inherently politically and ideologically loaded in vision and application, reshaping in particular ways how cities are managed and regulated. Likewise, the data within these systems are not neutral and objective in nature, but are situated, contingent and relational, framed by the ideas, techniques, technologies, people and contexts that conceive, produce, process, manage, analyse and store them (Bowker and Star, 1999; Lauriault, 2012; Ribes and Jackson, 2013). Drawing on our analysis of indicators, benchmarking and dashboards, we contend that the politics and technical and epistemological shortcomings of smart city initiatives need to be exposed and critiqued, not necessarily to call for them to be abandoned, but rather so they can be reimagined and repositioned in more inclusive, open and relational ways.

Five critiques of smart cities

There is a powerful political and economic lobby advocating the development of smart cities. The arguments forwarded by this lobby propose that smart city initiatives will lead to more efficient, effective, sustainable, resilient, safe and secure cities. This has been countered by critical scholars, policy analysts and community organisations, whose concerns can be divided into five broad themes: the growth of technocratic governance; the hollowing out of the state and the corporatisation of city governance; the creation of buggy, brittle and hackable city systems; the production of panoptic surveillance, predictive profiling and social sorting; and the promotion of an instrumental rationality and realist epistemology.

Technocratic governance

The first major concern about smart cities is that they adopt and promote technocratic forms of governance that presume that all aspects of a city can be measured, monitored and treated as technical problems that can be addressed through technical solutions. Such an approach is underpinned by what Mattern (2013) terms 'instrumental rationality' and practices what Morozov (2013) calls 'solutionism'. That is, there is a belief that complex open systems can be disassembled into neatly defined problems that can be solved or optimised through computation. All that is required to understand, manage and fix – in rational, logical and impartial ways – the issues a city faces is sufficient data and suitable algorithms. The critique of such an approach is threefold. First, a technocratic approach is highly reductionist and functionalist, always based on a limited selection of data and shaped by the formulation of algorithms, and fails to recognise the wider effects of culture, politics, policy, governance and capital in shaping city life and urban systems. Second, technological solutions largely focus on the efficient management of the manifestations of problems, rather than solving the deep-rooted structural problems underpinning them. As such, they largely paper over cracks rather than fixing them, unless coupled with a range of other policies. Third, technocratic control and command systems tend to centralise power and decision-making into a select set of administrative offices, rather than distributing power. Consequently, it is suggested that smart city initiatives will produce anaemic forms of top-down, centralised governance that do not live up to their promise.

Corporatisation of governance

The second concern is that, as well as being too technocratic in nature, the smart city agenda is being overly driven by corporate interests who are using it to capture government functions as new market opportunities (Greenfield, 2013; Townsend, 2013). Several of the world's largest digital technology and consulting companies operate smart city initiatives, including IBM, Cisco, Intel, Microsoft, SAP and Arup, and have become active players in city management, either through being key partners in building new smart cities from the ground up (e.g. Songdo or Masdar City), or partnering with established cities to retrofit their infrastructure with ICT and data solutions. While such companies might be fostering innovative and useful interventions there are three related anxieties concerning their foray into roles traditionally delivered by the state, especially those involving regulation and governance. First, it actively promotes a neoliberal political economy, the marketisation of public services and the hollowing out of the state, wherein city functions are administered for private profit (Hollands, 2008). Second, it potentially creates a technological lock-in or corporate path dependency that ties cities to particular technological platforms and vendors over a long period of time, creating monopoly positions (Hill, 2013). Third, it will lead to the creation of 'one-size-fits all smart-city-in-a-box' solutions that take little account of local cultures or political structures (Townsend *et al.*, 2011).

Buggy, brittle and hackable urban systems

The third major concern is that the ubiquitous use of digital technologies for running and managing city services and infrastructures is creating environments which are inherently buggy and brittle and are prone to viruses, glitches, crashes and security hacks (Kitchin and Dodge, 2011; Townsend, 2013). Technologies powered by software constitute an unusual product because they are sold in full knowledge that they are inherently partial, provisional, porous and open to failure. Such technologies routinely have to be patched and updated to cope with new contingencies. Further, they are vulnerable to being maliciously hacked, with the system subverted or shut down or valuable data stolen. As systems become ever more complicated, interconnected and dependent on software, producing stable, robust and secure devices and infrastructures becomes more of a challenge (Townsend, 2013). The notion of smart cities takes two open, highly complex and contingent systems – cities and digital systems – and binds them together. At the same time, these new systems lead to the discontinuation of analogue alternatives, meaning that if they fail there are no alternatives until the system is fixed/rebooted. The fear for some commentators is the creation of highly vulnerable and costly urban systems, rather than robust systems that create efficiencies and resilience.

Panoptic surveillance, predictive profiling and social sorting

Smart city technologies generate and are dependent on vast quantities of data. Many are the sources of what has been termed big data. That is, datasets that are: generated in real time; exhaustive rather than sampled; varied in nature; fine-grained in resolution; uniquely indexical in identification; relational; and flexible, holding the traits of extensionality (can add new fields easily) and scalability (can expand in size rapidly) (Kitchin, 2014b). Digital CCTV, retail checkout tills, smart phones, online transactions and interactions, sensors and scanners and social and locative media – produced by government agencies, mobile phone operators, app developers, internet companies, financial institutions, retail chains and surveillance and security firms – all generate massive amounts of detailed data about cities and their citizens. Such data are being routinely traded to and between data brokers as an increasingly important commodity, and examined by state security and policing agencies. For many commentators, the creation of smart cities raises questions concerning the creation of panoptic surveillance (gazing at the world) and wide-scale dataveillance (trawling through and interconnecting datasets), as well as anxieties relating to predictive profiling, social sorting and anticipatory governance that use data and algorithms to determine how people are treated (Kitchin, 2014b). The fear is that, far from being a liberatory and empowering development, smart cities may well lead to highly controlling and unequal societies in which rights to privacy, confidentiality, freedom of expression and life chances are restricted.

The politics of urban data

The final concern, and the focus of the rest of this chapter, is the politics of urban data. As already noted, the generation, processing and analysis of data is critical to

TABLE 2.1 The apparatus and elements of a data assemblage

Apparatus	Elements
Systems of thought	Modes of thinking, philosophies, theories, models, ideologies, rationalities, etc.
Forms of knowledge	Research texts, manuals, magazines, websites, experience, word of mouth, chat forums, etc.
Finance	Business models, investment, venture capital, grants, philanthropy, profit, etc.
Political economy	Policy, tax regimes, incentive instruments, public and political opinion, etc.
Governmentalities and legalities	Data standards, file formats, system requirements, protocols, regulations, laws, licensing, intellectual property regimes, ethical considerations, etc.
Materialities and infrastructures	Paper/pens, computers, digital devices, sensors, scanners, databases, networks, servers, buildings, etc.
Practices	Techniques, ways of doing, learned behaviours, scientific conventions, etc.
Organisations and institutions	Archives, corporations, consultants, manufacturers, retailers, government agencies, universities, conferences, clubs and societies, committees and boards, communities of practice, etc.
Subjectivities and communities	Of data producers, experts, curators, managers, analysts, scientists, politicians, users, citizens, etc.
Places	Labs, offices, field sites, data centres, server farms, business parks, etc., and their agglomerations
Marketplace	For data, its derivatives (e.g. text, tables, graphs, maps), analysts, analytic software, interpretations, etc.

Source: Kitchin (2014b: 25)

smart city initiatives. To be smart, that is to act with wisdom, one requires knowledge, which is dependent on information, which is extracted from data. As with smart city projects themselves, the data they rely on are portrayed as being objective and non-ideological. How can a sensor, a smartphone or a commercial transaction have a politics? They simply measure light or heat or app use or a trade – producing measurements and records that reflect the truth about the world. Data can thus be taken at face value. Likewise, it is argued, the algorithms used to process these data are neutral and non-ideological in their formulation and operation, grounded in scientific objectivity. Such a framing of data and algorithms enables smart city projects to present an image of being politically benign and commonsensical; that they help make a city secure, efficient, productive, sustainable and so on, by employing rigorous, technical practices that capture, process and analyse vast quantities of transparent, neutral, objective data.

Critics, however, contend that data are much more complicated in nature (Kitchin, 2014b). What data are generated is the product of choices and constraints,

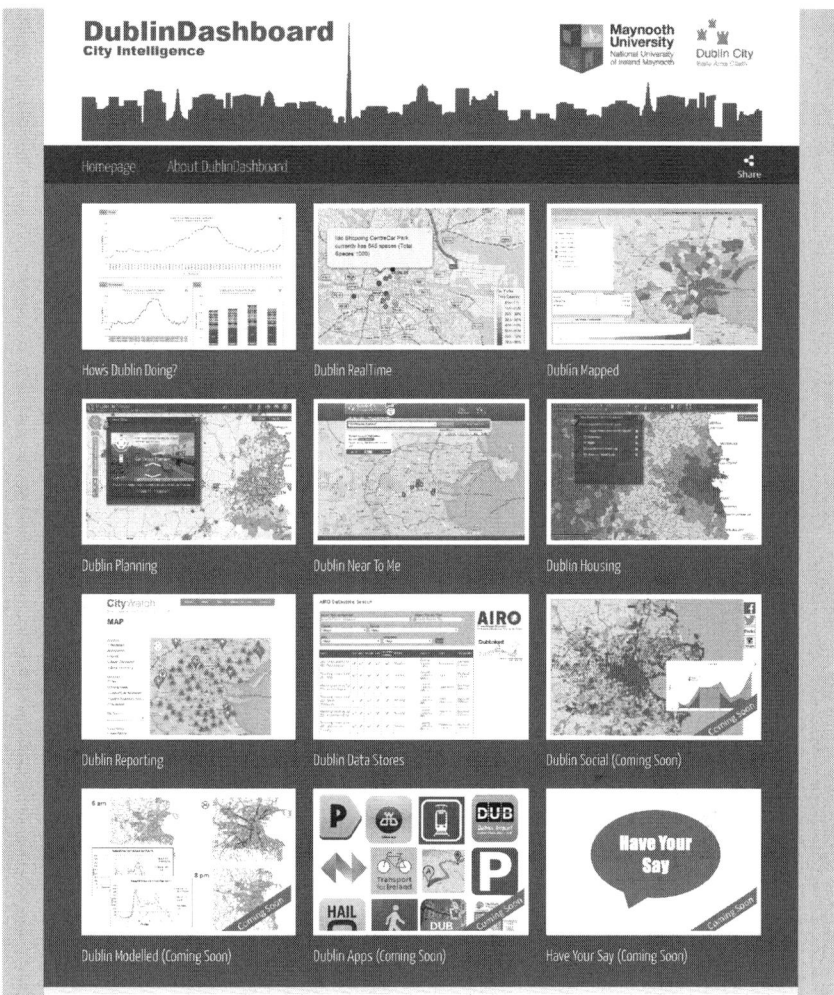

FIGURE 2.1 Dublin Dashboard
Source: www.dublindashboard.ie (used with permission)

shaped by a system of thought, technical know-how, public and political opinion, ethical considerations, the regulatory environment and funding and resourcing. What data are captured by a system is shaped by: the field of view/sampling frame (where data-capture devices are deployed, what their settings/parameters are, who uses a space or medium); the technology and platform used (such as different surveys, sensors, lenses, textual prompts and layout, which produce variances and biases in what data are generated); the context in which data are generated (unfolding events mean data are always situated and contextualised with respect to circumstance); the data ontology employed (how the data are calibrated and classified); and the regulatory environment with respect to privacy, data protection and security

(Kitchin, 2013). Data are situated, contingent, relational and framed and used contextually to try and achieve certain aims and goals. They are the product of a complex data assemblage (see Table 2.1).

This notion of a data assemblage is similar to Foucault's (1977: 194) concept of the 'dispositif' – a 'thoroughly heterogeneous ensemble consisting of discourses, institutions, architectural forms, regulatory decisions, laws, administrative measures, scientific statements, philosophical, moral and philanthropic propositions', which enhance and maintain the exercise of power within society. The dispositif of a data assemblage produces what Foucault terms 'power/knowledge', that is, knowledge that fulfils a strategic function. Such assemblage thinking, extended and reworked into a diverse set of post-structural and materialist mappings of both relational ontologies and the more flat ontologies of actor network theory, has been in recent years applied to urban studies (e.g. Farías and Bender, 2009; McGuirk and Dowling, 2009; McFarlane, 2011). Whatever the flavour of assemblage thinking, through such lenses it is clear that data are produced, managed, shared and deployed within heterogeneous ensembles which are never neutral, essential and objective in nature – data are never raw but always cooked to some recipe by chefs embedded within institutions that have certain aspirations and goals and operate within wider frameworks and constraints.

It is the tension between the realist epistemology (data show the city as it actually is) and the instrumental rationality of smart city systems, and an alternative view that exposes the politics and assemblages of such data and systems, that the rest of the chapter considers. It does so by critically examining urban indicator, city benchmarking and real-time dashboard projects, which are considered key municipal initiatives for enacting smarter governance. The analysis presented draws on an extensive reading of the literature and reflection on our own work building the Dublin Dashboard (see Figure 2.1), which blends indicators, benchmarks and dashboards into one extensive open-access data system for the city.

Urban indicators, city benchmarking and real-time dashboards

Urban indicators are recurrent quantified measures that can be tracked over time to provide a picture of stasis and change with respect to urban phenomena. A number of different indicator types can be deployed, which vary in their rationale and use. Single indicators consist of the measurement of a statistic related to a single phenomenon. Such indicators can be direct (e.g. measuring the phenomenon, such as R&D spend to reflect investment in innovation) or indirect (e.g. using a proxy, such as the number of patents registered) in nature. Composite indicators combine several single measures using a system of weights or statistics to create a new derived measure, recognising that most phenomena (e.g. social deprivation) are interrelated and multidimensional and that no one indicator can reveal the extent or complexities underpinning an issue (Maclaren, 1996). Single and composite indicators can be deployed in different ways as:

- *descriptive or contextual indicators* that provide overviews of phenomena and allow comparisons between locales that form inputs into policy formulation, but are not used in prescriptive or disciplining ways;
- *diagnostic, performance and target indicators* that are used to diagnose a particular issue or assess performance such as effectiveness (whether goals and objectives are being met – doing the right things) and efficiency (whether getting the most output for the input – doing things right and value for investment) of a policy or programme, individual workers, departments, organisations and sectors (Holden, 2006; Franceschini *et al.* 2007). Targets can be absolute (to reach a defined level) or relative (to match the performance of another organisation/place);
- *predictive and conditional indicators* that are not only considered good measures of present trends, but also key referents for predicting and simulating future scenarios and performances.

The most prized indicators are those that are considered well defined and unambiguous, are independent of external influence, have strong representativeness (they measure what they claim to measure), can be easily captured as a quantitative measure, are traceable over time, sensitive to change, verifiable and replicable, easy to interpret, timely (produced regularly and reported with minimal delay) and quick and cost-effective to collect, process and update (Franceschini *et al.*, 2007; Bhada and Hoornweg, 2009).

Urban indicator projects have proliferated since the early 1990s, driven by the sustainability goals of the United Nations Conference on Environment and Development (UNCED) in Rio de Janeiro in 1992 and in particular Chapter 40 of *Agenda 21*, and the rise of new managerialism and the desire to reform the public sector management of city services to make them more efficient, effective, and transparent and better value for money, combined with citizen and funder demands for evidence-based decision-making. As a consequence, cities around the world now routinely generate suites of indicator data, using them to track and trace performance, guide policy formulation and to inform how cities are governed and regulated. Indeed, in many locales, the use of urban indicators has become normalised as the *de facto* civic epistemology through which a public administration is measured and performance communicated (Miller, 2005). Further, indicators can be used as the inputs to urban models that seek to explain present patterns, and simulate and predict what might happen under different circumstances.

City benchmarking consists of comparing urban indicators within and across cities to establish how well an area/city is performing vis-à-vis other locales or against best practice. The process is often accompanied with score-carding, whereby tables of rankings and ratings, along with changes in relative position, are produced to reveal which places are doing well and who has caught up or fallen behind leading places (Gruppa and Mogee, 2004). Huggins (2009) details three types of area-based benchmarking: performance benchmarking that compares how well a place is doing with respect to a set of prescribed indicators; process benchmarking that compares the practices, structures and systems of places; policy benchmarking that compares public policies that influence performance and processes with respect to outcomes and meeting prescribed expectations. Benchmarking sets an aspirational and competitive agenda for cities and is used to motivate policy changes deemed necessary to alter a city's relative rating/ranking. When a city is relatively highly ranked, the scores are often also used in place-promotion to attract foreign direct investment and tourists. Jones Lang LaSalle report that there are now over 150 city benchmarking initiatives which seek to compare and contrast hundreds of cities (Moonen and Clark, 2013). In 2014, an ISO standard for city benchmarking indicators was announced (ISO, 2014a), designed to produce standardised, reputable, verifiable comparable global urban data (ISO, 2014b).

Dashboards visualise indicators through a graphic interface and are common sights in urban control rooms. Some dashboards seek to consolidate critical information onto a single screen using visualisation techniques such as gauges, traffic-light colours, meters, arrows, bar charts and graphs, which can be monitored at a glance (Few, 2006; see Figure 2.2). In contrast, analytical dashboards provide more extensive systems, acting as a console for navigating, drilling down into, visualising and making sense of numerous layers of interconnected data (Dubriwny and Rivards, 2004; see Figure 2.1). Dashboards are usually interactive, enabling users to interrogate and play with the data. Often, data can be simultaneously visualised in a number of ways, for example as a table, graph and map, with interactions in one pane being mirrored in the others. Dashboards thus provide a 'span of control' over a large repository of voluminous and varied data, and quickly transitioning data in the case of real-time data (Brath and Peters, 2004). As such they enable domains to be explored and interpreted without the need for specialist analytics skills (the systems are point and click and require no knowledge of how to produce graphics or maps). The power of dashboards is that they quickly and effectively provide city managers and to a lesser extent citizens with up-to-date detailed information about different aspects of urban systems and milieux, and how they are changing over time and space.

For their advocates, indicator, benchmarking and dashboard initiatives have high utility because they reveal in detail the state of play of cities. They enable a city to be known *as it actually is* and to assess how it is performing vis-à-vis targets and other places. They thus provide a powerful realist epistemology that shapes not only how cities are understood, but how cities are managed and governed. How such projects are translated into governance strategies, techniques and structures, however, varies between places.

For some municipalities, indicator, benchmarking and dashboard initiatives are being used to underpin forms of new managerialism. In such cases, they are used to guide operational practices with respect to specified targets, provide evidence of the success or failure of programmes and policies, discipline and reward performance and guide the development of new strategies (Craglia et al., 2004). Just as the dashboard of an aircraft cockpit provides detailed data about a plane and its flight, such projects are understood to provide city managers with data about the city and its management (Edwards and Thomas, 2005). Within such a framework, cities

Smart cities and the politics of urban data   25

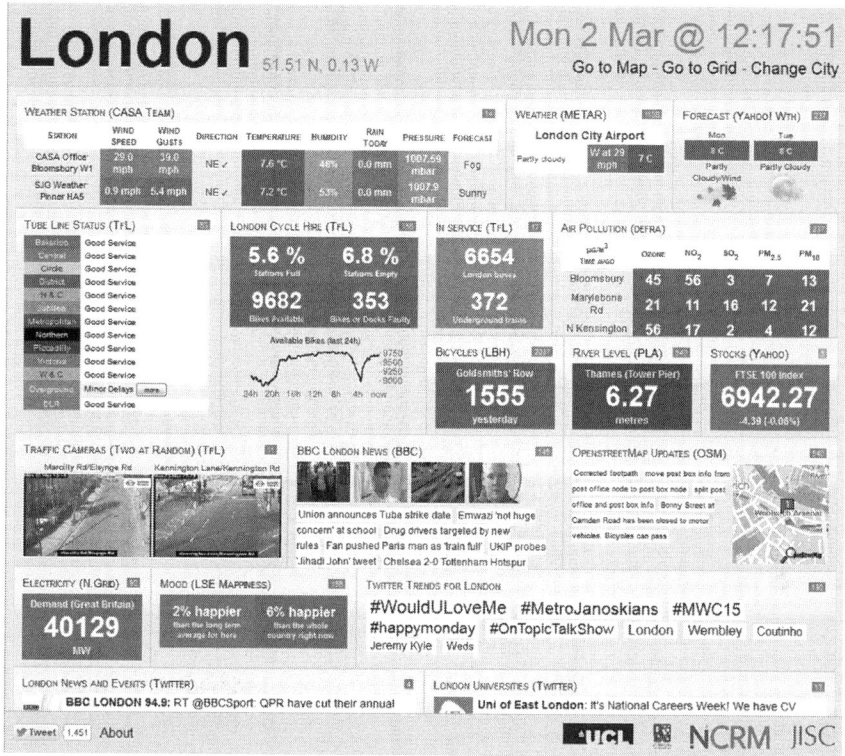

FIGURE 2.2   CASA London Dashboard
Source: citydashboard.org/london/ (used with permission)

are understood to consist of a set of knowable and manageable systems that act in 'rational, mechanical, linear and hierarchical' ways and 'can be steered and controlled with strong leadership, solid coordination, powerful (planning) instruments and/or high-quality guidance information' (Block and Van Assche, 2010: 3). There is thus a strong degree of instrumental rationality at play. An example of such an approach is Baltimore's use of CitiStat. Every week city managers meet in a specially designed room using dashboards to review performance and set new targets for the city as a whole and for each department (Gullino, 2009).

In contrast, some municipalities use indicators, benchmarking and dashboards in a more contextual way. Rather than cities being understood as mechanical systems that can be disassembled into their component parts and fixed, or steered and controlled through data levers, cities are conceived as consisting of multiple, complex, interdependent systems that influence each other in often unpredictable ways (Innes and Booher, 2000). Moreover, governance is seen as being complex and multi-level in nature, and the effects of policy measures are diverse and multifaceted, and neither is easily reducible to performance metrics and targets. Indicators highlight trends and potential issues, but do not show their causes or prescribe answers.

Deriving solutions to the issues facing cities then requires more than simply pulling levers in response to changes in indicator patterns. In such cases, indicators, benchmarking and dashboards are seen as one set of useful contextual data, but are not used in a strongly instrumentalist, mechanistic way to direct management practices. A long-standing example of such an approach is that employed within Flanders, Belgium, where since the late 1990s a number of cities have employed a common City Monitor for Sustainable Urban Development, consisting of nearly 200 indicators, to provide contextual evidence for policy-making (Block and Van Assche, 2010; Van Assche et al., 2010). The Dublin Dashboard follows this model.

In both new managerialism and contextual policy formulation, indicator, benchmark and dashboard projects form key initiatives in trying to implement more data-driven, evidence-based practices of governance and policy formulation. Their realist epistemology is favoured over more subjective and qualitative forms of information, because they provide objective, neutral facts that enable transparent, non-political, commonsensical policy- and decision-making. Not only does this provide better intelligence, it counters policy based on anecdote, cronyism and localism. However, such projects are far from being non-political, commonsensical and objective, as we now discuss.

The politics of indicator, benchmarking and dashboard initiatives

Indicator data supposes that facts can be abstracted from the world in value-free and objective ways and be benchmarked against each other. A fact is, after all, a fact and can be accurately measured – there are x number of people living in a city; x percentage of them are unemployed; there are x number of deaths from different illnesses; the trains are on average x minutes late, etc. In some cases, these facts can be measured using scientific instruments or algorithms mining databases; in others, by means of opinion surveys. In the latter case, it is assumed that a form of mechanical objectivity is deployed that adheres to defined rules and rigorous, systematic method to produce distant, detached, impartial and transparent data that are free of researcher bias and preferences, and are independent of local customs, culture, knowledge and context (Porter, 1995). Indicator and benchmark data can thus be accepted at face value as expressing a truth about the world (Kitchin, 2014a) and deployed through an instrumental rationality (Mattern, 2014).

As already noted, however, critics contend that a fact is never simply a fact. Facts are produced, not simply measured. How the rate of unemployment is calculated differs across locales (which one is 'true'?). It can often change within a jurisdiction, evolving through several iterations. Famously, the Thatcher Conservative government of the 1980s in Britain altered how unemployment was calculated twenty times in the space of a few years. When calculating how many people live in a city at any one time, who is selected for inclusion (e.g. are seasonal migrant workers included) and where is the boundary of a city (e.g. how much of its suburbs and hinterland is included)? Procedures and protocols, measurement instruments and scales, and standards used to generate facts are designed, negotiated and debated. Moreover, the production of facts is highly reductionist, atomising complex, contingent relationships into simplified, one-dimensional or composite measures. Reducing the city to a collection of facts decontextualises it from its history, its political economy, the wider set of social, economic and environmental relations that frame its development and its interconnections and interdependencies that stretch out over space and time (cities are open, not closed systems; Craglia *et al.*, 2004; Mori and Christodoulou, 2012). It thus produces a very particular, shallow representation of the city. Consequently, how data are ontologically defined as facts is not a neutral, technical process, but a normative, political and ethical one (Bowker and Star, 1999).

Further, such initiatives tend to gloss over technical issues that also undermine their supposed objectivity. As with all data, because indicator data are abstracted and representative there are always questions concerning data veracity and quality. These questions extend to how accurately (precision) and faithfully (fidelity) the data represent what they are meant to (especially when using samples and proxies), and how clean (error- and gap-free), untainted (bias free), consistent (few discrepancies) and reliable (the measurement instrument consistently produces the same quality of results) the data are (Goodchild, 2009; Kitchin, 2014a). The level of data trustworthiness also varies over time and place due to different measurement regimes and their evolution, as new technologies, practices and personnel are deployed (Ribes and Jackson, 2013). There are also ecological fallacy effects created through strategies such as aggregation. For example, indicator values often represent large, diverse areas, masking internal variation that can lead to false conclusions being drawn about a place as a result of how the underlying data are collated, categorised and presented. As a consequence, decisions over the statistical geography of a city can have a dramatic effect on indicator values and benchmarking ranking (Openshaw, 1984; Wrigley, 1995). Similarly, altering the relative weightings of data in composite indicators can have a profound effect on the resulting score. Indeed, many composite indicators are highly sensitive to adjustments and thus are vulnerable to manipulation, either through tinkering with the algorithm or by gaming the data (Gruppa and Mogee, 2004). Interpreting indicator data always requires, then, an appreciation of the level of uncertainty inherent in the data and analysis effects.

These various issues undermine the credibility of benchmarking initiatives which assume a universalism in the validity and standard of measures and method across place. Benchmarking also assumes that there is a normative standard by which places should be judged, some ideal state they are all seeking to achieve, rather than acknowledging that phenomena in different jurisdictions/places differ from one another often for good reasons (they have different aims, ambitions, histories, economies, etc.), and that how indicators relate to policy-making in one place may produce poor policy in another (Gruppa and Mogee, 2004; Huggins, 2009). Moreover, benchmarking is a zero-sum game in that cities are rated and ranked, with only one city being able to occupy each place, so that despite the fact that they may have improved their performance they may still be lowly ranked vis-à-vis other locales.

The use of dashboards has been an important factor in the promotion of indicator and benchmarking initiatives because they provide a powerful means to make sense of such data (through time and across space). Dashboards facilitate the illusion that it is possible to 'picture the totality of the urban domain', to translate the messiness and complexities of cities into rational, detailed, systematic, ordered forms of knowledge (Mattern, 2014: online). Dashboards, however, do not simply present the data. They too have a politics. Their makers might envisage them as detached, passive or neutral instruments that communicate the world as it is, but dashboards actively frame and do work in the world. They do not simply represent urban phenomena, but are constitutive of and actively produce meaning. As such, a dashboard seeks to act as a translator, not simply a mirror, setting the forms and parameters for how data are communicated and thus what the user can see and engage with. This translation is ideologically framed and inherently political, shaping what questions can be asked of the underlying data and what answers can be obtained.

Indicator, benchmarking and dashboard initiatives thus inherently express a normative notion about what should be measured and how it should be measured. They are full of values, judgements and deliberate strategies of occlusion. And the decisions taken have consequence for subsequent analysis, interpretation and action. Further, indicator, benchmarking and dashboard initiatives have a deep normative effect, being used to shape city governance, modify institutional behaviour, condition workers, influence decision-making and shape spending patterns (Franceschini et al., 2007). In this sense, they do not simply act as a camera reflecting the world as it is, but rather act as an engine shaping the world in diverse ways (MacKenzie, 2008). They not only represent urban systems, but actively help produce them. As Hezri (2004) details, indicators can be used in many different ways that are all politically charged: instrumentally (e.g. for problem-solving and decision-making); conceptually (e.g. to understand and interpret a situation); tactically (e.g. for delaying a strategy, to substitute for action, deflect criticism); symbolically (e.g. to provide reassurance or place promotion); and politically (e.g. as ammunition to support a particular position).

These politics and technicalities of indicator and benchmarking data and dashboards are framed and produced by their associated data assemblages. For example, with respect to the Dublin Dashboard, there were several internal team meetings and meetings with public officials concerning the purpose and principles of the dashboard, the selection of indicators and modules, obtaining access to data sources, dashboard design, selection of software, roll-out and use of the system, future maintenance and so on. Each issue was debated and consensus established, framed within the wider context of the systems of thought, ideologies and communities of practice of the developers and city officials; their knowledge base of data relating to the city, the handling, analysing and displaying of such data and coding skills; the financial constraints of the project and the wider political economy, policy regime, governmentalities and legalities; the materials and infrastructures available to them; and so on (see Table 2.1). In other words, there was nothing inevitable about the scope, form, and operation of the Dublin Dashboard, nor the qualities of the data presented through it; instead they emerged through design, tinkering, debate and negotiation between stakeholders, framed with in a wider assemblage of ideas, institutions, policies, regulations, laws, finance, etc.

Conclusion

Since 2010 and the launch of IBM's Smarter Cities challenge, there has been a growing call from business, government and academia for the creation of smart cities. This call seeks to build on the roll-out of networked urbanism and the embedding of digital technologies into the fabric of cities over the next couple of decades. The result is a powerful discursive regime that promotes the vision of smart urbanism – cities that seek to leverage digital technologies to produce secure, efficient, productive, competitive, sustainable, resilient urban locales. The smart city is forwarded as the most effective way of coping with the projected enormous growth in urban populations, adapting to climate change and other environmental shocks and enabling efficiencies whilst coping with shrinking public budgets. For the most part, the creation of smart cities is presented as a pragmatic, non-ideological, commonsensical approach to dealing with the various issues facing cities. However, as we have discussed in this chapter, such a claim is disingenuous and there are a number of unresolved concerns about the development of smart cities. These include anxieties related to the rise in technocratic governance, the corporatisation of governance, the creation of buggy, brittle and hackable systems, panoptic surveillance, predictive profiling and social sorting and the politics of urban data.

This chapter has largely concentrated on examining the latter by considering the roll-out of indicator, benchmark and dashboard initiatives, which constitute one set of smart city technologies. These initiatives purport to provide detailed city intelligence, including real-time overviews of how the city is performing. Their power is derived from a realist epistemology that claims to show the city as it actually is, along with an instrumental rationality that translates factual information into actionable knowledge. As a consequence, such initiatives have quickly become key technologies in how many cities are managed and governed, though how they are conceived and deployed can vary quite markedly. We have sought to trouble this epistemology and rationality by exposing their politics and technical issues, demonstrating that their claims to 'truth' are little more than claims. Instead, such initiatives are framed by and within wider data assemblages and are plagued by technical and methodological conundrums. They are the outcome of normative concerns and they have normative effects.

Nonetheless, we are of the view that indicator and dashboard projects have utility, providing information of value to city managers and citizens. Such initiatives provide valuable spatially extensive and time-series data about the state of play of cities. They provide an evidence base far superior to anecdote, and have advantages over one-off studies in terms of coverage and regularity. Rather than being abandoned, instead we contend they need to be reimagined and positioned, openly recognising and acknowledging: (i) the multiple, complex, interdependent nature of cities, which means that they cannot be simply disassembled into a collection of facts; (ii) that indicators and dashboards do not merely reflect cities, but actively frame and produce them; (iii) that they are not mechanistic toolkits but data assemblages – complex socio-technical systems infused with politics and context; and (iv) that there are multitude other ways to see and understand the city that produce valuable, insightful knowledge.

It may well be the case that those who develop and promote indicator, benchmark and dashboard initiatives are already aware of their contingencies, relationalities, politics and technical shortcomings, but deal with them by engaging in a form of strategic essentialism that covers them over or pretends they do not exist in order to promote their approach (and, in the case of industry, products) and to deflect possible critique. If this is the case, then we are advocating that the fig-leaf of such strategic essentialism be tugged away. The stakes with regards to how cities are managed and governed using such initiatives are too high – particularly when they are used to direct resources and formulate and implement policy. This is the approach we have taken in the Dublin Dashboard initiative, seeking to be reflexive as to how the project is framed, understood and practised within an assemblage of various actors and actants.

In our view such reimagining and repositioning needs to occur across smart city technologies. While they might be pragmatic approaches to the myriad issues facing cities, none of them is non-ideological and commonsensical. Instead they are all infused with politics and technical issues that need to be recognised and acknowledged. It is through such a strategy that more emancipatory and empowering visions of smart cities can be developed that best serve the common good and not simply the market ambitions of companies or the control desires of states.

Acknowledgements

The research for this chapter was funded by a European Research Council Advanced Investigator award (ERC-2012-AdG-323636-SOFTCITY) and by Science Foundation Ireland.

References

Bhada, P. and Hoornweg, D. (2009) *The Global City Indicators Program: A More Credible Voice for Cities*. Washington, DC: World Bank, Urban Development Unit.

Block, T. and Van Assche, J. (2010) Disentangling urban sustainability: the Flemish City Monitor acknowledges complexity. Paper presented at the Seventh International Conference on Ecological Informatics: Unravelling Complexity and Supporting Sustainability, Ghent, Belgium, December 2010. Available at: https://biblio.ugent.be/input/download?func=downloadFile&recordOId=1090655&fileOId=1090661 [Accessed 17 July 2014).

Bowker, G. and Star, L. (1999) *Sorting Things Out: Classification and Its Consequences*. Cambridge, MA: MIT Press.

Brath, R. and Peters, M. (2004) Dashboard design: why design is important. *DM Direct*, October 2004.

Caragliu, A., Del Bo, C. and Nijkamp, P. (2009) *Smart Cities in Europe*. Research Memoranda Series 0048. Amsterdam: VU University Amsterdam, Faculty of Economics, Business Administration and Econometrics.

Craglia, M., Leontidou, L., Nuvolati, G. and Schweikart, J. (2004) Towards the development of quality of life indicators in the 'digital' city. *Environment and Planning B* 31 (1): 51–64.

Datta, A. (2015) New urban utopias of postcolonial India: 'entrepreneurial urbanization' in Dholera smart city, Gujarat. *Dialogues in Human Geography* 5 (1): 3–22.

Dubriwny, D. and Rivards, K. (2004) Are you drowning in BI reports? Using analytical dashboards to cut through the clutter. *DM Review*, April 2004. Available at: www.advizorsolutions.com/press/Cut%20Through%20The%20Clutter.pdf [Accessed 4 June 2014].

Edwards, D. and Thomas, J. C. (2005) Developing a municipal performance-measurement system: reflections on the Atlanta Dashboard. *Public Administration Review* 65 (3): 369–376.

Farías, I. and Bender, T. (eds) (2009) *Urban Assemblages: How Actor-Network Theory Changes Urban Studies*. London: Routledge.

Few, S. (2006) *Information Dashboard Design: The Effective Visual Communication of Data*. North Sebastopol, CA: O'Reilly.

Foucault, M. (1977). The confession of the flesh. In *Power/Knowledge: Selected Interviews and Other Writings, 1972–1977*, ed. C. Gordon. New York: Pantheon Books, 1980, pp. 194–228.

Franceschini, F., Galetto, M. and Maisano, D. (2007) *Management by Measurement: Designing Key Indicators and Performance Measurement Systems*. Berlin: Springer.

Goodchild, M. F. (2009) Uncertainty. In R. Kitchin and N. Thrift (eds) *International Encyclopedia of Human Geography*, vol. 12. Oxford: Elsevier, pp. 1–5.

Greenfield, A. (2006) *Everyware: The Dawning Age of Ubiquitous Computing*. Boston: New Riders.

Greenfield, A. (2013) *Against the Smart City*. New York: Do Projects.

Gruppa, H. and Mogee, M. E. (2004) Indicators for national science and technology policy: how robust are composite indicators? *Research Policy* 33 (9): 1373–1384.

Gullino, S. (2009) Urban regeneration and democratization of information access: CitiStat experience in Baltimore. *Journal of Environmental Management* 90: 2012–2019.

Hezri, A. A. (2004) Sustainability indicators system and policy processes in Malaysia: a framework for utilisation and learning. *Journal of Environmental Management* 73 (4): 357–371.

Hill, D. (2013). On the smart city: or, a 'manifesto' for smart citizens instead. *City of Sound* [online], 1 February. Available at: www.cityofsound.com/blog/2013/02/on-the-smart-city-a-callfor-smart-citizens-instead.html [Accessed 5 February 2013].

Holden, M. (2006) Urban indicators and the integrative ideals of cities. *Cities* 23 (3): 170–183.

Hollands, R. G. (2008) Will the real smart city please stand up? Intelligent, progressive or entrepreneurial? *City* 12 (3): 303–320.

Huggins, R. (2009) Regional competitive intelligence: benchmarking and policy-making. *Regional Studies* 44 (5): 639–658.

Innes, J. and Booher, D. E. (2000) Indicators for sustainable communities: a strategy building on complexity theory and distributed intelligence. *Planning Theory & Practice* 1 (2): 173–186.

ISO (2014a) ISO 37120 Sustainable development of communities: indicators for city services and quality of life [online]. Available at: www.iso.org/obp/ui/#iso:std:iso:37120:ed-1:v1:en [Accessed 8 October 2014].

ISO (2014b) How does your city compare to others? New ISO standard to measure up [online]. Available at: www.iso.org/iso/home/news_index/news_archive/news.htm?refid=Ref1848 [Accessed 17 July 2014].

Kitchin, R. (2013) Big data and human geography: opportunities, challenges and risks. *Dialogues in Human Geography* 3 (3): 262–267.

Kitchin, R. (2014a) The real-time city? Big data and smart urbanism. *GeoJournal* 79 (1): 1–14.

Kitchin, R. (2014b) *The Data Revolution: Big Data, Open Data, Data Infrastructures and Their Consequences.* London: Sage.

Kitchin, R. and Dodge, M. (2011) *Code/Space: Software and Everyday Life.* Cambridge, MA: MIT Press.

Lauriault, T. P. (2012) Data, infrastructures and geographical imaginations: mapping data access discourses in Canada. Unpublished PhD thesis, Carleton University, Ottawa.

McFarlane, C. (2011) The city as assemblage: dwelling and urban space. *Environment and Planning D: Society and Space* 29 (4): 649–671.

McGuirk, P. M. and Dowling, R. (2009) Neoliberal privatisation? Remapping the public and the private in Sydney's masterplanned residential estates. *Political Geography*, 28 (3): 174–185.

MacKenzie, D. (2008) *An Engine, Not a Camera. How Financial Models Shape Markets.* Cambridge, MA: MIT Press.

Maclaren, V. W. (1996) Urban sustainability reporting. *Journal of the American Planning Association* 62 (2): 184–202.

Mattern, S. (2013) Methodolatry and the art of measure: the new wave of urban data science [online]. *Design Observer: Places*, 5 November. Available at: https://places-journal.org/article/methodolatry-and-the-art-of-measure/38174/ [Accessed 15 November 2013].

Mattern, S. (2014) Interfacing urban intelligence [online]. *Places: Design Observer*, April. Available at: http://places.designobserver.com/feature/how-do-we-interface-with-smart-cities/38443/ [Accessed 17 July 2014].

Miller, C. A. (2005) New civic epistemologies of quantification: making sense of indicators of local and global sustainability. *Science, Technology and Human Values* 30 (3): 403–432.

Moonen, T. and Clark, G. (2013) *The Business of Cities 2013: What Do 150 City Indexes and Benchmarking Studies Tell Us about the Urban World in 2013?* Chicago: Jones Lang LaSalle. Available at: www.jll.com/Research/jll-city-indices-november-2013.pdf [Accessed 17 July 2014].

Mori, K. and Christodoulou, A. (2012) Review of sustainability indices and indicators: towards a new City Sustainability Index (CSI). *Environmental Impact Assessment Review* 32: 94–106.

Morozov, E. (2013) *To Save Everything, Click Here: Technology, Solutionism, and the Urge to Fix Problems That Don't Exist.* New York: Allen Lane.

Openshaw, S. (1984) *The Modifiable Areal Unit Problem.* Concepts and Techniques in Modern Geography 38. Norwich: Geo Books.

Porter, T. M. (1995) *Trust in Numbers: The Pursuit of Objectivity in Science and Public Life.* Princeton, NJ: Princeton University Press.

Ribes, D. and Jackson, S. J. (2013) Data bite man: the work of sustaining long-term study. In L. Gitelman (ed.) *'Raw Data' Is an Oxymoron.* Cambridge, MA: MIT Press, pp 147–166.

Townsend, A. (2013) *Smart Cities: Big Data, Civic Hackers, and the Quest for a New Utopia.* New York: W.W. Norton.

Townsend, A., Maguire, R., Liebhold, M. and Crawford, M. (2011) *A Planet of Civic Laboratories: The Future of Cities, Information and Inclusion.* Palo Alto, CA: Institute for the Future.

Van Assche, J., Block, T. and Reynaert, H. (2010) Can community indicators live up to their expectations? The case of the Flemish City Monitor for Livable and Sustainable Urban Development. *Applied Research Quality Life* 5: 341–352.

Vanolo, A. (2014) Smartmentality: the smart city as disciplinary strategy. *Urban Studies* 51 (5): 883–898.

Wrigley, N. (1995). Revisiting the modifiable areal unit problem and the ecological fallacy. In A. D. Cliff, P. R. Gould, A. G. Hoare and N. J. Thrift (eds) *Diffusing Geography: Essays for Peter Haggett.* Oxford: Blackwell, pp. 49–71.

3

IBM AND THE VISUAL FORMATION OF SMART CITIES

Donald McNeill

Introduction

> What is decisive is that from the 19th Century the government of the city becomes *inseparable from the continuous activity of generating truths about the city* … Urban thought here is technical and practical, not simply dreams of the city but mundane techniques of gathering, organisation, classification, and publication of information.
>
> *(Osborne and Rose, 1999: 739)*

> Albuquerque, New Mexico, is using a business intelligence solution to automate data sharing among its 7,000 employees in more than 20 departments, so every employee gets a single version of the truth. It has realized cost savings of almost 2,000 percent.
>
> *(IBM, 2014: online)*

This chapter starts from the observation that a recurrent theme in the smart cities discourse is a reference to truth, objectivity, evidence, along with other similar concepts aligned with the positivist tradition. And that there exists a 'data' of cities that has become a self-evident truth, and one that requires addressing: from its collection, to its analysis, to its output in the form of a generated truth. This 'big data' turn is, of course, a key element of contemporary commercial, urban and social policy in many sectors, from finance to education, health to media (Kitchin, 2014).

This has put an onus on the leadership of municipal governments, especially those that cover the territories of large, complex cities, to make sense of the 'deep torrent of timely, varied, resolute and relational data' (Kitchin, 2014: xv) that it gathers and stores as a central part of its statutory existence. Understanding the significance of these 'mundane techniques' is often not in the skill set of local government department heads, more accustomed to a career path based upon the management of large public sector labour forces and budgets. This gap has been filled by global technology firms such as IBM, who, I argue, have been at the forefront of naturalising 'single versions of the truth'.

Central to this process has been the development of visual technologies that provide several different modes of visual perception of a city's domains of governance. My argument here is that these technologies have been central to the visual formation of smart cities, both in an ontological sense (that cities which cannot be seen in such ways are by definition not smart) and in a practical sense (that cities that cannot be viewed cannot be made to work in a smart manner). This requires placing 'smart cities' within a longer genealogy of city government rather than seeing them as part of some new digital era.

To pursue this idea, this chapter draws particularly closely from two influential studies of nineteenth-century visual practice: Chris Otter's *The Victorian Eye: A Political History of Light and Vision in Britain, 1800–1910* (2008) and Jonathan Crary's *Suspension of Perception* (2001). These studies can be read productively, I suggest, as providing a set of parallel perspectives on how these ways of interpreting the constitution of nineteenth-century British urban government can be applied to contemporary studies of smart cities. To take this one step further, it could be argued that only through practices of visualisation are state operatives able to perform acts of government. This chapter provides an elaboration of this in respects ways: first, through allowing introspection, enhancing the state's ability to 'see itself'; second, as a means of grasping the city as a totality through synopsis; third, through practices of supervision, particularly in allowing its operatives to maintain the quality and consistency of the repetitive practices that underpin it; fourth, through practices of foresight, where the city's leaders are able to speak confidently of its envisioned futures. Following a brief exposition of IBM's work in this area, I expand the discussion of what is meant by these terms.

IBM, data visualisation and the perception of cities

> City employees in white jumpsuits work quietly in front of a wall of giant screens – a sort of virtual Rio, rendered in real time. Video streams in from subway stations and major intersections. A sophisticated weather program predicts rainfall across the city. A map glows with the locations of car accidents, power failures, and other problems.
>
> *(Singer, 2012: online, writing for the* New York Times*)*

In 2010, Rio de Janeiro opened an operations centre (Figure 3.1). This city, facing the challenge of hosting both the football World Cup and the Olympics in very close proximity, and notorious for its crime, gridlock and poor municipal planning, had engaged IBM as part of its decision to restructure its municipal service provision. The invitation had been prompted, according to its mayor Eduardo Paes, by the

FIGURE 3.1 Rio de Janeiro's Center of Operations, COR
Source: Andrés Luque-Ayala

severe storm which hit Rio earlier in the year, which led to significant landslides in the hillside favelas, flash flooding and the inundation of key highways, and which caused the deaths of 68 people. It became clear to Paes that the city had no location from which the mayor could *oversee* rescue operations: in other words, find a privileged point from which the city could be visualised in a synoptic manner and then acted upon. Shortly thereafter, he invited an IBM team to Rio; they responded by identifying the need to bring together the diverse sets of data held by different government departments into a single dataset, which could then be analysed from a central control point. As a solution to the inundations that had caused such misery in the city previously, IBM trialled its beta weather forecasting tool, Deep Thunder, which brought together historical weather data and topographical analysis in a way which sought to provide advance warning to city government of an inundation. Seizing upon Rio's most iconic event, the operations centre was also configured to coordinate the complex logistics of the city's *Carnaval* street events, allowing departments to 'assign time slots to the street bands and map their routes, as well as plan for security, street cleaning, crowd control and other needs' (Singer, 2012: online).

Interestingly, IBM's Rio team did not identify the need for a new building that would act as a control centre. Their focus was on analytical and visualisation software, data acquisition and optimising the hardware of the city's computing networks. And this has reflected the changing nature of IBM as a firm: their interest in Rio was a significant manifestation of the shift in direction of one of the world's most famous corporations. In 2002, the company had announced that it would divest itself of the parts of the firm which had for many years been the base of its success, its computer and business hardware design and manufacturing sections. Instead, it would focus itself more narrowly on the field known as business services, which at that time was being revolutionised by new developments in enterprise software. The intention, IBM announced, was to 'move up the value chain' into 'more lucrative fields'. The company launched a suite of vertical specialisms under the rubric of a 'Smarter World'; these included health, energy, marketing, and financial services. Yet within these specialisms it also identified a range of horizontal initiatives, such as a 'Smarter Cities' area, which it quickly brandmarked (Söderström et al., 2014).

Launched in 2010, IBM's Smarter Planet campaign (which itself was both a marketing campaign and a way to organise and valorise the firm's global engineering and human capital capabilities) gave city services a high priority. This interest in cities, followed also by several global competitors such as Cisco and Siemens, requires a simplification and standardisation of the problems they face, in order to provide an economically scalable set of solutions (McNeill, in press). Beginning with a series of showcase global cities, such as Rio and Singapore, and a large number of pilot projects in relatively lesser-known cities, from Dubuque to Fremantle, the firm has elaborated a distinct way of representing or 'telling stories' about cities (Söderström et al., 2014). Snapshots of these can be viewed via the firm's YouTube channel, and are also elaborated – in general terms – in a series of Redguides that provide the broad parameters of their approach (e.g. IBM, 2010a, 2010b).

The focus of this discussion is a single product: IBM's Intelligent Operations Center. Launched in 2012, this software package is designed to run on established computer networks. IBM describes the business-value case for the product as follows:

- Helps city officials better monitor and manage city services by providing them insight into daily city operations through centralized management and data intelligence.
- Helps city agencies prepare for problems before they arise and to coordinate and manage problems when they do arise.
- Enables officials to communicate instantly and discuss and synchronize rescue efforts so they can send the correct people and equipment to the correct places at the correct times.
- Facilitates cross agency decision making, convergence of domains, coordination of events, communication, and collaboration, which improves the quality of services to the citizens and reduces expenses.
- Flags event conflicts automatically between city agencies.
- Optimizes planned and unplanned operations using a holistic reporting and monitoring approach.
- Helps operations executive or staff to adjust systems to achieve results that are based on the insights gained.

(IBM, 2012: 4)

The most striking feature of the product is its dashboard. This screen-based user interface provides a set of different, animated images which are concurrently present on a single screen. While these can be customised, they may include: two-dimensional maps of a city district with key event locations marked on; live-streamed photographic images of the area from a mounted camera; a graph that plots the changing relationship between, for example, rainfall and water level in a flood-affected area; and various other headline statistics which are enlarged to allow easy and immediate comprehension.

Through the creation of a dashboard, IBM's most striking contribution was its ability to visualise the data that it had gathered and analysed. This apparently panoptical instrument was demonstrated by Mayor Paes in a TED Talk in 2012:

> For the climax, he turned to the screen and dialed up a videoconference with Carlos Roberto Osorio, his point man for urban affairs, back in Rio. For the next minute, Osorio flipped through a dizzying succession of live digital maps and debriefed the mayor on the day's events – the GPS-tracked movements of the city's garbage truck fleet, current precipitation picked up by the city's brand-new Doppler radar, and Deep Thunder's latest forecast (all clear).
>
> *(Townsend, 2013: 68)*

The 'visual workspace' of the package is thus a key element of the Intelligent Operations Center for Smarter Cities. As the guide puts it:

> The IBM Intelligent Operations Center user interface is a dashboard that provides insight into data that is customized to a user's role and authority. This flexible view into the wealth of data that is flowing into, and stored in IBM Intelligent Operations Center, is at the heart of the solution. Its appearance is configurable and delivers exactly the data the user wants to see and is allowed to see.
>
> *(IBM, 2012: 9)*

The question that this chapter now addresses is as follows: what is the data that the typical user, a city executive, wants to see? And how is it incorporated in various practices of city governance? We might at this point draw on Crary's (2001) study of attention, and consider that 'Attention as a process of selection necessarily meant that perception was an activity of *exclusion*, of rendering parts of a perceptual field unperceived' (24–5; original emphasis). In other words, while enhanced visuality is central to the smart city moment, it is only screening out certain elements that allows visual interpretations to be operationalised as a workable practice of city leadership. The following section describes how this takes place.

Introspection: the city-state gains self-knowledge

Vision and visuality has long been held to be a core element of urban governance, especially in the social reform agenda of many late nineteenth-century city governments (Joyce, 2003). However, it is in Chris Otter's *The Victorian Eye* that we find this fully discussed. As Otter elaborates, the emergence of regimes of visualisation has been an important element in the governmentality of cities. He goes as far as to identify 'nine basic, recurring patterns of visual perception: oligoptic oversight, supervision, inspection, privacy, obfuscation, voyeurism, distant and simple signification, proximate and complex signification, and detail' (2008: 256). Otter's array ranges from the visual signifier of the town hall clock to a set of human gazes and practices, but he is quick to point out that his taxonomy is 'emphatically non-discrete. None operated as entirely self-enclosed perceptual paradigms or scopic regimes, especially since they invariably overlaid one another, producing complicated, unpredictable configurations of visual experience' (256). However, what is important here is that the state played a key role in stitching together these varied visual domains and practices into some degree of order.

This work fits within a wider set of studies of how the state gains knowledge of itself, a topic which has been given some significant attention in recent years (e.g. Hannah, 2010; Mitchell, 2002; Joyce, 2003; Scott, 1999). In this corpus of work, there is an interest in examining the work of the state in enabling itself to act, particularly through the use of devices of categorisation, information gathering and measurement. This is often implicated in a capitalistic 'will to improve', which, as Tania Murray Li (2007: 7) has argued, involves a dual process of problematisation ('identifying deficiencies that need to be rectified') and 'rendering technical' (which 'constitutes the boundary between those ... with the capacity to diagnose deficiencies in others, and those who are subject to expert direction'). This has been accompanied by a significantly more detailed approach to the state than has often been the case in earlier traditions of, particularly Marxian, analysis. It is, as Osborne and Rose (1999) argue, an important stage in the immanence of the city.

In this context, the assembly of things becomes the state's basic role: in their work on Paris, Latour and Hermant (2006) sets out to specify and locate the small number of places in the city where government takes place. These are usually offices of some kind or other. It is interesting that Latour has developed this approach from his study of scientific laboratories, because he has something of an obsession with the offices of professional knowledge brokers, which he refers to as 'centers of calculation'. He offers a 'Wall Street trading room' or 'Bill Gates' as exemplars: he introduces two key terms: 'oligopticon' and 'panorama'. Both refer to modes of vision and knowing. The former is important because of its quality: 'From oligoptica, sturdy but extremely narrow views of the (connected) whole are made possible – as long as connections hold' (Latour, 2005: 181). The latter he uses to gather together those moments when someone tries to make sense of 'the whole situation' or the 'Big Picture', things that 'nicely solve the question of staging the totality' (188), whether it be a textbook, a scientist summarising the field of science for a lay audience or a CEO addressing a shareholder meeting.

From this perspective, one of the key elements of smart city data visualisation is *introspection*: the ability of the state to look within and open its own black boxes where key data is hidden or locked from scrutiny by departmental heads. One of the best known, and notorious, of these examples came during Rudy Giuliani's period as mayor of New York, during which he used an early 'smart city' technology, Compstat, as a means of targeting policing resources. His reflections on his time in office in this interview, seen through Crary's thesis of attention, are instructive:

> Well, I very much subscribe to the 'Broken Windows' theory, a theory that was developed by Professors Wilson and Kelling, 25 years ago maybe. The idea of it is that you had to pay attention to small things, otherwise they would get out of control and become much worse. And that, in fact, in a lot of our approach to crime, quality of life, social programs, we were allowing small things to get worse rather than dealing with them at the earliest possible stage … So we started paying attention to the things that were being ignored. Aggressive panhandling, the squeegee operators that would come up to your car and wash the window of your car whether you wanted it or not – and sometimes smashed people's cars or tires or windows – the street-level drug-dealing; the prostitution; the graffiti, all these things that were deteriorating the city. So we said, 'We're going to pay attention to that,' and it worked.
>
> *(Giuliani, 2003: online)*

This form of attentiveness was one of a number of strategies used by Giuliani during his time as mayor (McNamara and McNeill, 2012). Ward police chiefs were periodically grilled if their local crime statistics were not satisfactory: the immediate impact of this mode of attention was felt in the streets, disproportionately by African-American youth.

And so perhaps the key disciplining, introspective force of the dashboard is the display of Key Performance Indicators (KPIs), which have become central elements in new managerial discourse within the public sector. Despite the firm's stress on the complexity of data, the IBM product does the opposite, creating a singular data point and colour-coding it:

> KPIs are used to measure nearly anything of importance to city leaders, from the number of traffic accidents this calendar quarter to the on-time performance of the public transportation system. IBM Intelligent Operations Center receives raw or computed metrics and uses them to compute the actual KPI. For example, for bus performance, the metrics might indicate, for each bus, whether it is ahead of schedule, on time, or behind schedule. When rolled up with all the other bus information, it might compute to a single metric that indicates whether, on average, the buses are on schedule. City bus administrators can rest easy if they see, at one glance, that the average bus arrival is green. This status probably means that, on average, buses are arriving at approximately their scheduled times. If this KPI turns yellow or red, the administrator can determine the cause and act appropriately.
>
> *(IBM, 2012: 12)*

In this mode of visual perception, then, the hierarchies of urban government are cascaded onto the screen through the simplicity of colour-coded KPIs. These KPIs thus become enrolled in the disciplining process of government. From then on, one assumes, a downward pressure will be placed on public sector employees by the dashboard. Despite benign promises of good government, such data dashboards are a core part of what Kitchin (2014: 61–2) identifies as the 'neoliberalisation and marketisation of public services', linked with a New Public Management ethos.

Synopsis: the visualisation of the totality

While introspection and 'paying attention' to urban problems can easily lead to the scapegoating of particular sections of the urban population, or public sector workforce, the state also seeks to grasp its total existence. For smart city enthusiasts, this is the great promise of 'big data'. If it could be said that the only way that cities can actually exist in an ontological sense is through a visual representation of their totality, then data visualisation probably affords that better than ever before. The creation of a particular version of a totality, though, is a political and moral act. As John Law puts it:

> Knowledge practices, and the forms of knowledge that these carry, become sustainable only if they are successfully able to manage two simultaneous tasks. First, they need to be able to create knowledge (theories, data, whatever) that work, that somehow or other hold together, that are convincing and (crucial this) do whatever job is set for them. But then secondly and counter-intuitively, they have to be able to generate realities that are fit for that knowledge.
>
> *(Law, 1986: 240)*

In a similar vein, Otter's (2008: 259) core argument is that the knee-jerk fear of 'panopticon cities' diverts attention from the very real oligoptical regimes that govern cities – that there are only a small number of groups powerful enough to exert power over a city's population using technology. Importantly, though, Otter's argument is based less on the 'macrological, statist, and legalistic approach to liberalism' and more on individual agency, on the 'very local, technological textures of visual practice':

> At this microscale, the operation of power is often better captured through the idioms of norm and capacity … We are speaking here of the agency made possible by technological networks. The numerous, interlaced vision networks or patterns stimulated and sustained a panoply of individual vision norms and capacities: productive attention, sensory awareness, urban motility, social observation, private reading.
>
> *(Otter, 2008: 259)*

In this perspective, a characteristic of modern governance is that road users, for example, are fully enmeshed in a network of sociality, sensation and consciousness held together by socio-technical practices of driving, car ownership and so on. This provides an alternative urban paradigm to that of the technocratic agglomeration of individuality that Law (2009) has discussed as being the ontological definition of the mass survey.

IBM uses this synoptic approach in its attempt to rewrite the concept of the city centre, where the historically central city origins are less important than having a system-centric vantage point. Here, one of the firm's executives explains via a video demonstration how the Operations Center works:

> The last place we sent that information about the event was to 'city central'. Now, why 'city central' again? Well, if we get a water overflow issue – and in this case it affected transportation as well, the buses – the city should know about it. Over on the left-hand side, this is the key performance indicator scorecard I mentioned before, and we can see each of the departments – if they're reaching their target, what their actual results are – or I should say what is their target, what are their actual results? We see in orange here that we've got a problem with both 'waste management' and in this case 'bus' – the one we've been working on. If I look up here there is a 'city disruption'. So I, in the mayor's office, in city central, am going to take a look at this; and the system said, 'Hey, there is a potential water overflow issue; a potential water overflow issue; and it's going to impact food distribution.' Well, this certainly should become a priority for the city. So 'city central' creates a 'directive'. By creating a directive, up here it says, 'This is a Sev 1. This is severity 1 and I want this to be on the top of the list for things that get serviced by the individual departments – priority one.' So I'm going to go ahead and submit that directive and now it's gone off to that individual department and said, 'Make it the top priority on your list.' We'll fast-forward in time here – boom – and select that 'city directive' again. We can see now that the water overflow issue has been resolved. I can go from 'city central', close the 'directive' and when I do, 'green light – go'; we've fixed the problem. As you can see, by instrumenting and interconnecting the systems within a city and making the systems more intelligent, we can make a smarter city and we can make a smarter planet.
>
> (IBM Smarter Cities Operations Center, n.d.)

It is interesting, in the context of urban history, to think about this in tandem with that other signifier and enabler of centrality, the town or city hall. City central, in the IBM discourse, does not place as much significance on the building as on a more mobile, flexible site of knowledge:

> We assert that, in a smarter city, information in the form of metrics, events, and processes must be shared across organizations in a near real-time manner.

> In a smarter city, city-wide operational processes using data from any number of domains can continuously predict and react to events and trends that are affecting the city. Taking action leads to rebalancing and, therefore, optimization. Optimization must include two dimensions: both the goals of the individual domains and those of the city as a whole.
>
> *(IBM, 2010a: 7)*

Indeed, it is worth noting that the smart city discourse, at this point in its history, is often used in the context of exploration and discovery, a point that could be tied to the discourse of cities as laboratories (Karvonen and van Heur, 2014). This in turn has strong discursive linkage with microscopes and the technology of examining the miniature.

Such an approach has been given detailed consideration by Latour and Hermant in *Paris, ville invisible* (2006), a sociological travelogue which seeks to establish the ontological basis of Paris composed using the principles of actor network theory. A key point here is how social theory has long felt able to effortlessly ignore the nature of the size of a particular social entity:

> All too often social theory still inhabits this utopian world where the zoom is possible. It really believes that we can slide from biggest to smallest, and then wonders how the microscopic – face-to-face interaction – manages to remain meaningful despite the crushing weight of the macroscopic.
>
> *(Latour and Hermant, 2006: 59)*

As *Paris, ville invisible* develops, it becomes strikingly clear that Latour and Hermant's journey is about inserting themselves in the ontological 'flatlands' of Paris. They are seeking out as many apparently powerful sites as possible, and make the journey there to meet, observe and note the humans and objects that together fuse to create the city as we think we know it. It would be quicker, cheaper and analytically more satisfying to make general statements about how 'government controls traffic', but the point is that we don't really know this, we only assume it. For example, they find themselves in the 'control' room of the Parisian traffic police, standing before multiple screens which stream images from various points on the city's ring road to the controller, a Mr Henry. Their point here is to show that actor network theory is not about eliminating such obvious 'god-like' viewing stations (describing Mr Henry as 'the missing figure of the panopticon' (Latour and Hermant, 2006: 51)) but rather to contrast the fact that the flattening of the academic analyst's vision is not the same as flattening the actor's vision. The point here, which Latour had elaborated in *Reassembling the Social* (2005), is that at some point, the explanatory power of social relations fails to prove a causation or make a case, and turns to how an individual might be figured:

> You don't have to imagine a 'wholesale' human having intentionality, making rational calculations, feeling responsible for his sins, or agonizing over his mortal soul. Rather, you realize that to obtain 'complete' human actors, you

have to compose them out of many successive layers, each of which is empirically distinct from the next.

(Latour, 2005: 207)

Without getting into the very vital debates about the utility of actor network theory, it is worth considering the often casually grandiose claims made by smart city advocates: the ability to grasp the 'whole' city and its interrelated problems. This kind of reading can sometimes be subject to a 'surface' interpretation: 'As soon as we follow the shifting representation of the social we find offices, corridors, instruments, files, rows, alignments, teams, vans, precautions, watchfulness, attention, warnings – not Society' (Latour and Hermant, 2006: 17). The core message of *Paris, ville invisible* is that the city can be captured as a unity, but only provisionally, and with hard work. Smart city dashboards claim to do the same, but their confidence in claiming to have confidently found 'city central' is misplaced.

Supervising and inspecting urban infrastructure

It has recently been suggested that insufficient attention is paid among scholars to the ordering of repetitive things and maintenance regimes. As Graham and Thrift (2007) suggest, this is a crucial, but understudied, element of how capitalism perpetuates itself. A materialist analysis of how the city state makes inventories of objects, for example, is thus important. Carter *et al.* (2011: 7) provide the example of how street furniture, such as bins, bollards and benches, 'are understood as minor "actants" in the city yet microcosms that contain the whole city and are witnesses to the forces, wills, relations, subjectivities and power that make it up'. These apparently mundane technologies (and the ability of engineers to attach sensors or cameras to them) are in fact key props in the development of smart cities. And while introspection and synopsis are more generalised processes of observation of the object of the city, the smart city has become associated with new modes of instrumentation and sensing with their own modes of governance (e.g. Gabrys, 2007; see also Chapter 6). In the nineteenth century, citizens were less blasé about this: a whole set of now mundane municipal practices such as street-cleaning and restaurant inspection grew up as the state moved into supervisory mode. And so:

> The consensual, negotiated, and legally governed nature of the [inspection] system reflected liberal beliefs in personal privacy and the rule of law … regimes of inspection left large areas of private, individual existence altogether uninspected, or, in a quite formal sense, free: central or local government had a very clear sense that it had a right to see only so much.
>
> (Otter, 2008: 133)

The nineteenth-century city government became concerned with hygiene, above all: overcrowding of tenements, the condition of animals and their connection to human consumption, the condition of pipes, the safety of infrastructure.

The legacy of these Victorian era infrastructures can be seen in cities like New York. One interesting intersection between these legacies and the smart city is offered by Mayer-Schönberger and Cukier (2013: 68–70), who discuss the case of New York's electricity network manhole covers. These pieces of urban technology have historically had a tendency to overheat and at times explode. In the 2000s, a team from Columbia University analysed the data held by Con Edison in an attempt to optimise the company's ability to identify priority maintenance cases. This brought forth the classic problem faced in using old 'small data' holdings. For example, the standard component known as the 'service box' was referred to in no less than 38 different ways in the handwritten notes of maintenance engineers (from SB through SB/X through SERV-BOX).

And so, unsurprisingly, smart cities vendors have been interested in the marketability of smart sensors in the home, government and business as a means of altering both how energy is consumed and how pipes and switches are maintained. This poses important issues of measurement. Otter's study of Victorian governmentality included a consideration of how – particularly gas – light was measured in comparison with oil- or candlelight. In recounting some of the controversies of the time, Otter points out that there were

> several early-nineteenth-century epistemological conundrums surrounding illumination. When one measured light, what exactly was one measuring? What instruments should be used? What should be the unit of measurement? How bright should streetlights be?
>
> *(Otter, 2008: 136)*

He continues: 'In 1809, there was no defined standard against which to measure gaslight, no fixed unit of measurement, and no consensus about the kind of apparatus one should use to make the comparison.' For example, to establish gaslight as a system, governments and companies had to create both a physical apparatus to carry the gas to point of use, and also 'numerous regulating, monitoring, inspection, and recording processes necessary to maintain its smooth government. These regulatory agents (photometrists, inspectors, meters, governors) were sometimes rather inefficient, even ramshackle, but they aimed to subject both mains and light to an acceptable degree of control and predictability' (Otter, 2008: 136).

We might link this to IBM's interest in putting smart (computer-readable) sensors on manhole covers and pipes: the smart city's eyes are thus entirely systematised (Lohr, 2009). This is revealed in the re-emergence of cybernetic thinking within IBM's methodology, as Söderström *et al.* have argued:

> There is something apparently odd in this resurrection as it gives the audience of the smarter cities campaign a sense of traveling back to the heroic times of post-war cybernetics. If we consider urban dynamics as a translation device used for the purpose of storytelling, this choice becomes less enigmatic. What urban systems theory provides, seen from this perspective,

> is primarily a powerful metaphor creating a surface of equivalence. It translates very different urban phenomena into data that can be related together according to a classical systemic approach which identifies elements, interconnections, purposes, feedback loops, delays, etc. Thus, the website is packed with schemes and flash animations showing how contemporary cities are constituted by functioning and measurable (but highly perfectible) urban systems and infrastructures.
>
> *(Söderström et al., 2014: 313)*

This supervisory mode can be read as part of a benevolent technocracy. Here, the smart city has no interest in the individual's thoughts and practices, only that the city's resources are run efficiently and well. In the nineteenth century, this was a key part of liberal ideology: 'Inspection was designed to be thorough but spatially circumscribed and unintrusive: the inspector examined one's water mains or cows, not one's conscience, political beliefs, religious proclivities, or morals' (Otter, 2008: 133). In the contemporary city, it is interesting to chart debates about how far smart cities become wrapped up in illiberal political projects, where inspection and supervision of everyday practice is linked to moral judgements about individuals' lives and beliefs.

Foresight: envisioning the future city

> How are future[s] made present, whether that be through specific affects such as fear or anxiety, materialities such as development reports, or epistemic objects such as climate change graphs or inflation predictions?
>
> *(Anderson and Adey, 2012: 1533)*

> IBM believes that big data has the potential to significantly change how organisations use data and run analytics. Business analytics and big data are strategic bets for IBM, which recognizes the huge potential to demonstrate leadership in this space and create shareholder value.
>
> *(Parasuraman, 2012: online)*

According to Krishnan Parasuraman, CTO of digital media and general business, IBM has invested $16 billion across 30 acquisitions of start-up firms doing analytics, such as Netezza and Vivisimo. Combined with the world's largest patent portfolio and biggest commercial research background, it has produced a big data platform covering 'unstructured data management, text analytics, image feature extraction, and large-scale data processing' (Parasuraman, 2012: online).

For this reason, IBM might well be pioneering what we could call the 'stochastic city' – the construction of the city as an unstable object through the use of algorithms, visualisation and modelling. The firm has undertaken two high-profile examples where they have sought to demonstrate this in practice relating to weather and traffic. In Rio, IBM have suggested that through the analysis of hyperlocal weather patterns they can produce a correlation of weather events to predict things such as major floods, and thus allow for prompt evacuation of residents. In terms of modelling traffic flow, the company has argued that it can use predictive techniques to foresee future road space trends. Bernie Meyerson, IBM's VP of Innovation, has argued that through the study of the history of car transport data in Singapore, it could predict the time and location of traffic jams:

> We know from history what happens in Singapore if you slow the lights down in one direction by three seconds, and how to tweak the model so the jam never happens ... And so there will be a traffic jam that never occurs because we can predict what happens 20 minutes from now, because we can take enough Big Data and crunch it, and do analytics on it. So we're predicting the future, and changing it.
>
> *(Bernie Meyerson, in Gertner, 2012: 140–1).*

This raises some interesting questions about the return of urban modelling within city planning, which was adopted with enthusiasm during the 1960s and 1970s as the possibilities offered by mainframe computing were explored. The vastly enhanced capabilities of contemporary computing have allowed IBM to enter this field, part of a wider move towards a 'cybernetics redux' (Townsend, 2013). IBM, adapting their systems modelling knowledge from industrial consultancies, approached the city of Portland – a medium-sized city about to set out a 25-year plan – about jointly developing a city simulation. However,

> the spiderweb of equations ... quickly ballooned to over seven thousand equations (a number that was deemed too complex), was pruned back to six hundred (too simple), and then eventually built back up to the roughly three thousand contained in the final revision.
>
> *(Townsend, 2013: 83).*

The principal IBM strategist leading the Portland simulation, Justin Cook, pointed out to Townsend that the advent of web interfaces and digital platforms has changed the context of simulation modelling, shifting from a behind-the-scenes laboratory model of acting upon citizens, to one where – ideally – the model is co-produced and partially crowdsourced.

Indeed, it could be the very precarious nature of local government planning that makes digital modelling of the type that large technology firms offer so appealing. Harvey (2009) describes the development of a 3D urban model of Manchester produced by Arup, and points out that it has three distinct roles in urban governance, particularly in mediating the 'dynamic tensions between description, prediction and communication' (Harvey, 2009: 262). Thus it is important not to reify the power of such visually based models:

> From the inside plans and models look more contingent – and it is remarkable how often they are revisited, revised and re-drawn. On the most basic level design and construction engineers know that the physical environment is unstable, that conditions change, that averages are imprecise, and that imprecision is inevitable. That is why the business of measurement, calculation, analysis and prediction is so central to their work. These processes are repeated over and over again as the works proceed, and plans are continually tested, checked and refined. They know that samples generate statistical probabilities, not descriptive certainties.
>
> *(Harvey, 2009: 262)*

This insight is far-reaching, as it certifies that absolute knowledge of how a city will perform can never be known: only a set of possible scenarios can be produced. In this case, city leaders are not being sold certainty about the future, but are rather being sold a privileged vantage point from which to view various possible futures, and to decide – on the basis of risk, personal, electoral or financial – whether or not to take anticipatory action.

Conclusion

> The city might have been a text, but nobody would ever read it from cover to cover. Attention creates islands of detail, foregrounds and backgrounds, dividing social language from streams of noise or nonsense. Focus, scrutiny, discernment: all are little techniques of sifting and managing the flux of sensory data.
>
> *(Otter, 2008: 50)*

This chapter has explored what I have termed the visual formation of smart cities, in which several modes of visual perception have been built into, and marketed into, smart city software packages. It has identified some linkages between technopolitical practices of the urbanising state, particularly in the Victorian period of urban municipalism, and the smart cities strategies pursued by IBM and other corporations. From IBM's perspective, of course, it is merely enabling public managers to operate effectively in a hostile fiscal environment. But here the value-free, dehumanised technical systems approach also operates to justify a minimum, or socio-spatially targeted, service delivery:

> The Smarter City's time has come. Most cities already collect massive amounts of data, it just needs to be put to better use. Investing in information technology is one of the most fiscally prudent options for many municipalities that face severe service cuts. These are indeed difficult times, but they force us to re-imagine the future of cities. From where we're sitting, it looks pretty bright.
>
> *(IBM, 2013: online)*

This apparently casual use of perspective ('from where we're sitting') thus contains an invitation to city governments to abandon some of the more negative perceptions of urban futures (with their loaded metaphors of gloom, murk and darkness) and embrace a new set of visualisation tools. The visually enabled agent, it is implied, is more able to adjust to 'the future of cities'.

For this reason, it is important to conclude by considering how the range of modes of 'smart' visual perception – introspection, synopsis, supervision/inspection and foresight – might be applied to the study of political leadership. In his analysis of the speeches of Tony Blair, for example, Norman Fairclough draws attention to the often rapid shifts Blair made between 'what is the case (epistemic modality), what will be (predictions), what should be (deontic modalities), yet on the other hand speaking personally ("I" statements) and on behalf of an inclusive "we"' (Fairclough, 2005: 181). I have noted what happened when early smart thinking was applied in Giuliani's New York via Compstat. When we consider how a big-city mayor such as Giuliani might combine such visual tools with well-honed discursive practices into a political performance, we might get a sense of how smart city technology will be transported away from the apparently mundane and value-free norms of good city government (McNamara and McNeill, 2012). It is easy to make the case using synopsis that 'we' would be better off by using big data to profile certain undesirable minorities; that by using a narrow set of foresight criteria, it is better to cut certain social services to avoid future indebtedness; that inspection and supervision at a household level could be used illiberally by an extremist government or city council; and that the introspective view of city government might lead to the downward targeting of certain social programmes as being fiscally irrational. Here the everyday spatial engineering opportunities offered by smart technologies may not be as extreme as the 'desk killing' undertaken by Nazi operatives such as Walter Christaller (Barnes and Minca, 2013), but there are potentially darker futures than the bright horizons envisioned and arranged by IBM.

References

Anderson, B. (2010) Preemption, precaution, preparedness: anticipatory action and future geographies. *Progress in Human Geography* 34 (6): 777–798.

Anderson, B. and Adey, P. (2012) Future geographies. *Environment and Planning A* 44: 1529–1535.

Barnes, T. and Minca, C. (2013) Nazi spatial theory: the dark geographies of Carl Schmitt and Walter Christaller. *Annals of the Association of American Geographers* 103 (3): 669–687.

Carter, S., Dodsworth, F., Ruppert, E. and Watson, S. (2011) *Thinking Cities through Objects*. CRESC Working Paper 96. Milton Keynes: Centre for Research on Socio-Cultural Change (CRESC), Faculty of Social Sciences, The Open University.

Crary, J. (2001) *Suspensions of Perception: Attention, Spectacle, and Modernity*. Cambridge, MA: MIT Press.

Fairclough, N. (2005) *Analysing Discourse: Textual Analysis for Social Research*. London: Routledge.

Gabrys, J. (2007) Automatic sensation: environmental sensors in the digital city. *The Senses and Society* 2 (2): 189–200.

Gertner, J. (2012) Calling Dr Watson. *Fast Company*, November, 124–129, 140–141.

Graham, S. and Thrift, N. (2007) Out of order: understanding repair and maintenance. *Theory, Culture and Society* 24 (3): 1–25.

Giuliani, R. (2003) Interview: Rudolph Giuliani, Former Mayor of New York City [online]. Academy of Achievement. Available at: www.achievement.org/autodoc/printmember/giu0int-1 [Accessed 21 May 2015].

Hannah, M. (2010) *Dark Territory in the Information Age: Learning from the West German Census Boycotts of the 1980s*. Aldershot: Ashgate.

Harvey, P. (2009) Between narrative and number: the case of Arup's 3D digital city model. *Cultural Sociology* 3 (2): 257–276.

IBM (2010a) A Foundation for Understanding IBM Smarter Cities. Available at: www.redbooks.ibm.com/redpapers/pdfs/redp4733.pdf [Accessed 7 February 2014].

IBM (2010b) Introducing the IBM City Operations and Management Solution. Available at: www.redbooks.ibm.com/redpapers/pdfs/redp4734.pdf [Accessed 7 February 2014].

IBM (2012) IBM Intelligent Operations Center for Smarter Cities Administration Guide. Available at: www.redbooks.ibm.com/redbooks/pdfs/sg248061.pdf [Accessed 7 February 2014].

IBM (2013) Sponsorship statement, World Cities Summit, Singapore [online]. Available at: www.worldcities.com.sg/ibm_2012.php [Accessed 7 February 2014].

IBM (2014) Smarter cities: overview [online] Available at: www.ibm.com/smarterplanet/ae/en/smarter_cities/overview/ [Accessed 7 February 2014].

IBM Smarter Cities Operations Center (n.d.) Executive presentation, transcribed from a video uploaded to YouTube by the user IBM Open Standards (21 October 2010). Available at: www.youtube.com/watch?v=BT3Q4zyiNfA [Accessed 29 December 2012].

Joyce, P. (2003) *The Rule of Freedom: Liberalism and the Modern City*. London: Verso.

Karvonen, A. and van Heur, B. (2014) Urban laboratories: experiments in reworking cities. *International Journal of Urban and Regional Research* 38 (2): 379–392.

Kitchin, R. (2011) The programmable city. *Environment and Planning B: Planning and Design* 38: 945–951.

Kitchin, R. (2014) *The Data Revolution: Big Data, Open Data, Data Infrastructures and their Consequences*. London: Sage.

Latour, B. (2005) *Reassembling the Social: An Introduction to Actor-Network-Theory*. Oxford: Oxford University Press.

Latour, B. and Hermant, E. (2006) *Paris, Invisible City* [online] (first published in 1998 as *Paris, ville invisible*, Paris: La Découverte). English translation by Liz Carey-Libbrecht. Available at: www.bruno-latour.fr/sites/default/files/downloads/viii_paris-city-gb.pdf [Accessed 28 February 2015].

Law, J. (1986) On the methods of long-distance control: vessels, navigation, and the Portuguese route to India. In J. Law (ed.) *Power, Action and Belief: A New Sociology of Knowledge?* Sociological Review Monograph 32. London: Routledge and Kegan Paul, pp. 234–263.

Law, J. (2009) Seeing like a survey. *Cultural Sociology* 3 (2): 239–256.

Lohr, S. (2009) Bringing efficiency to the infrastructure [online]. *New York Times*, 30 April. Available at: www.nytimes.com/2009/04/30/business/energy-environment/30smart.html?pagewanted=all&_r=0 [Accessed 25 September 2012].

McNamara, K. and McNeill, D. (2012) The city personified: the geopolitical narratives of Rudy Giuliani. *Communication and Critical/Cultural Studies* 9 (3): 259–278.

Mayer-Schönberger, V. and Cukier, K. (2013) *Big Data: A Revolution That Will Transform How We Live, Work, and Think*. Boston: Eamon Dolan.

Mitchell, T. (2002) *Rule of Experts: Egypt, Techno-Politics, Modernity*. Berkeley: University of California Press.

Murray Li, T. (2007) *The Will to Improve: Governmentality, Development, and the Practice of Politics*. Durham, NC: Duke University Press.

McNeill, D. (in press) Global firms and smart technologies: IBM and the reduction of cities. *Transactions of the Institute of British Geographers*.

Osborne, T. and Rose, N. (1999) Governing cities: notes on the spatialisation of virtue. *Environment and Planning D: Society and Space* 17: 737–60.

Otter, C. (2008) *The Victorian Eye: A Political History of Light and Vision in Britain, 1800–1910*. Chicago: University of Chicago Press.

Parasuraman, K. (2012) IBM's strategy for big data and analytics [online]. The Big Data and Analytics Hub. Available at: www.ibmbigdatahub.com/blog/part-iii-ibm%E2%80%99s-strategy-big-data-and-analytics [Accessed 20 May 2015].

Pryke, M. (2010) Money's eyes: the visual preparation of financial markets. *Economy and Society* 39 (4): 427–459.

Scott, J. C. (1999) *Seeing like a State: How Certain Schemes to Improve the Human Condition Have Failed*. New Haven: Yale University Press.

Singer, N. (2012) Mission control, built for cities [online]. *New York Times*, 3 March. Available at: www.nytimes.com/2012/03/04/business/ibm-takes-smarter-cities-concept-to-rio-de-janeiro.html [Accessed 25 September 2012].

Söderström, O., Paasche, T. and Klauser, F. (2014) Smart cities as corporate storytelling. *City* 18 (3): 307–320.

Townsend, A. M. (2013) *Smarter Cities: Big Data, Civic Hackers and the Quest for a New Utopia*. New York: W.W. Norton.

4

THE SMART ENTREPRENEURIAL CITY

Dholera and 100 other utopias in India

Ayona Datta

100 smart cities, 100 utopias

In May 2014, the newly elected Prime Minister of India announced an ambitious national programme for building 100 new smart cities in India. Justified on the basis of rising global challenges of rural–urban migration and rapid urbanisation, this programme was initiated to transform India's urbanisation in line with global urban initiatives on smart urbanism. Following Townsend's broad definition, the 100 Indian smart cities aspire to become 'places where information technology is combined with infrastructure, architecture, everyday objects and our own bodies to address social, economic and environmental problems' (Townsend, 2013: 15). The 100 smart cities programme, although still largely in its conceptual stages, has stoked the Indian imagination for a potential shift in state–citizen relationships – from a perceptibly corrupt and opaque bureaucracy to an imagined democratic space of digital governance. In this condition, high-tech digital interventions in housing, infrastructure and governance as envisaged in smart cities are also perceived as the route to wider democracy in India.

My argument in this chapter, then, is that India's smart urbanism is a hybrid, a home-grown story of a global 'smart urbanism' which manifests in the 'smart entrepreneurial city'. To this end Indian smart cities draw upon what Jessop and Sum (2000) call 'the narratives of enterprise' to present the smart city as a space for middling local entrepreneurs who would become the new smart citizens. India's smart urbanism is synonymous with the rhetoric of innovation and enterprise, while, which, standing for ambiguous claims to modernity and development, are essentially measured by their contribution in pushing up India's GDP and widening India's footprint in the global economy. The question remains (following Luque-Ayala and Marvin, 2015) – how, for whom and to what consequences is the smart city being produced in India?

To answer this question, I subject the algorithmic urbanism of India's smart cities to scrutiny not by examining their claims to a digital utopia, rather by examining the state's coercive politics in materialising smart cities. I ask how the imaginaries of smart cities are being used to drive aspirations for modernity and development and to what socio-political and material outcomes. This approach, I believe, is essential since it reveals that India's smart urbanism is inherently contingent upon elementary processes of land accumulation by dispossession. Using the case of the alleged first smart city in India, Dholera, I will suggest that it is 'a powerful tool for the production of docile subjects and mechanisms of political legitimisation' (Vanolo, 2014: 883). While a rhetoric of crisis around migration and urbanisation in India legitimises the 100 smart cities programme, the case of Dholera outlines a city made and unmade through its own spatio-temporality. Speed and slowness are essential features of Dholera and by extension of smart urbanism in India – speed is produced by the urgency of rhetoric and policies while slowness emerges from the struggles for land, livelihoods and homes which the smart city attempts to erase.

I begin therefore by 'provincialising' the smart city in India in order to highlight a particularly popular trend of entrepreneurial urbanisation which constructs Dholera and other proposed smart cities as sites of economic growth. I then discuss the politics in transforming actually existing Dholera – a constructed terrain of *terra nullis* – into the image of a 'smart entrepreneurial city'. Following the speed of law- and policy-making that can materialise Dholera and its corresponding 'roadblocks' of local resistance action, I conclude that land is the final frontier of smart urbanism in India. The extraction of surplus value from 'unproductive' land and its transformation into real estate is the new face of the 'crisis' of urbanisation and hence of smart urbanism in India.

'Provincialising' the smart city in India

Dholera, a new 'smart city' built from scratch about 100 km from Ahmedabad, is key to India's smart city initiatives. As conceived, it is expected to cost around $9–10 billion, with the Indian state and Japanese corporations (Hitachi, Mitsubishi Corporation, Toshiba, JGC and Tokyo Electric Power Company) contributing up to 10 per cent of this amount; the rest is expected to come from the private sector. Presented as the first smart city in India in the State of the Union Budget in 2013 – although several other cities also make claim to this position now – and again in 2014 as part of the 100 smart cities programme, Dholera can be understood as an 'invented Eden' (Kargon and Mollela, 2008). It is an assemblage of tropes on smart urbanism (intelligent, sentient, real-time and networked cities), global urbanism (eco-cities, walkable cities, new urbanism and so on) and Indian urban planning (satellite cities, industrial townships, knowledge cities, finance cities, special economic zones) that seeks to mirror society's understanding of current technologies and at the same time 'regain the lost virtues of village life without sacrificing the undoubted gains of industrial advance' (Kargon and Mollela, 2008: 13). Through an 'importation of off-the-shelf program techniques' (Peck, 2002: 344) Dholera's masterplanning is completed by the UK-based global consultancy firm Halcrow and its 'smart' credentials are marketed by Cisco as a meshwork of fibre-optic cables, sensors and cameras linked to a central control room to track city-wide utility consumption.

In charting a critical agenda for smart urbanism, Luque-Ayala and Marvin (2015: 6) note that 'understanding the politics of the implementation of smart requires exploring how the smart city is constituted discursively, techno-materially and spatially'. Certainly Dholera smart city resonates with wider global urbanism trends such as gated communities, new towns, satellite cities and other spatial manifestations that scholars have labelled elsewhere as the 'Dubaisation of Africa' (Choplin and Franck, 2010), 'worlding' of cities (Roy and Ong, 2011) and 'assemblage urbanism' (McFarlane, 2011). Historically it also resonates with the techno-cities of the twentieth century – 'planned city developments in conjunction with large industrial or technological enterprises, blending the technological and the pastoral, the mill town and the garden city' (Kargon and Mollela, 2008). Dholera's smart urbanism is one that transcends ideologies and spans national borders, producing a new urban colonialism in a city of premium networked spaces (Graham, 2000) where urban planning as well as management and control of data will serve the interests and aspirations of the political elite and middle classes (Choe et al., 2008).

At another level, though, Dholera is not a 'new' city typology *per se*. It is an extension of the rhetoric and practices of 'new townships' and 'satellite cities' that have driven postcolonial urban experiments in India. As part of the 100 smart cities programme Dholera produces 100 'urban fantasies' propagating 'the hope that these new cities and developments will be "self-contained" and able to insulate themselves from the "disorder" and "chaos" of the existing cities' (Watson, 2014: 15). As a smart city built from scratch, Dholera can be seen to extend the focus of a neoliberal state on global cities (such as Mumbai), knowledge cities (such as Ambani City), technology cities (such as HITEC City), IT hubs (such as Bangalore), eco-cities (such as Lavasa), finance cities (such as GIFT) and so on, to a more digitally led city-making initiative in recent years. It reflects the almost perpetual presence of an 'investor-friendly' urbanism, which is legitimised through a 'metaphysics of disorder' (Comaroff and Comaroff, 2006) that is seen to define existing Indian cities in colonial and postcolonial periods.

I suggest that it is important to 'provincialise' Dholera and the 100 smart cities programme as the site of intense local and regional politics that presents a 'mutation' (Rapoport, 2014) of the globally circulating smart city model in India. In doing so, I align myself with Chakrabarty's (1992: 328) suggestion of 'developing the problematic of non-metropolitan histories' by unpacking and making visible the 'repression and violence that are as instrumental in the victory of the modern as is the persuasive power of its rhetorical strategies' (340). This means not just 'identifying and empowering new loci of enunciation' (Sheppard et al., 2013: 895) for situating the story of smart cities in the regional state of Gujarat, but also unpacking the 'ambivalences, contradictions, the use of force, and the tragedies and ironies'

(Chakrabarty, 1992: 340) associated with its vision to create 100 smart cities on an 'entrepreneurial model' in India. As elaborated in the recent smart city concept note released by the Government of India,

> Smart Cities are those that are able to attract investments. Good quality infrastructure, simple and transparent online processes that make it easy to establish an enterprise and run it efficiently are important features of an investor friendly city. Availability of the required skills in the labour force and adequate availability of electricity, water, etc. are important features of a Smart City. Investors, themselves, are looking for a decent living and so they also look for housing, healthcare, entertainment and education. Safety and security are as important to them as to any other resident.
>
> *(Government of India, 2014)*

The 100 smart cities programme in India provincialises the global trope of smart urbanism through the construction of a 'smart entrepreneurial city' and several other hybrid city tropes, such as 'smart industrial' (e.g. Dholera), 'smart finance' (e.g. GIFT), 'smart ecocity' (e.g. Lavasa), 'smart heritage' (e.g. Varanasi) cities, which appropriate existing cities within a global trope of the smart city. The instruments that are used to realise this are based on 'clean' technologies in infrastructure, participation of the private sector and e-governance. Yet the focus on big data and central surveillance and control that are so symptomatic of global smart cities such as Barcelona, Boston or Rio are muted here by a mere suggestion of the 'use of ICT' that means full Wi-Fi coverage across the city as well as connected homes and buildings. This is perhaps due to the fact that existing Indian cities must now bid for 'smart city' status by claiming a part of the Rs 700 crore total budget allocated for this initiative. This roughly translates to £7m per city, which is wholly inadequate to finance any 'smart city' initiatives without the active involvement of the private sector. Indeed a recent survey (Infotech, 2014) of local authority budgets showed that existing cities have allocated up to 1 per cent of their budget to ICT infrastructure; therefore it will be a big challenge creating smart cities, where in this budget will have to rise substantially.

Dholera can be provincialised by seeing its links with a postcolonial legacy of 'new township' building as a route to modernity and development in India. Similar to postcolonial Indian experiments with modernity in Chandigarh, Bhubaneshwar and Gandhinagar, Dholera, too, is arguably a 'blueprint utopia' (Holston, 1989) that has been designed to bring in a new era of social and economic prosperity in the regional state of Gujarat and beyond. It is, however, also distinctly different from earlier utopian experiments in one significant way. As a smart city it is driven not by visionary architects and planners but rather by the corporate sector seeking to create new global markets in India (IBM, 2010). While Dholera might be one of the first smart cities built from scratch in India, it will certainly not be the last. What is important to note here is that although Dholera predates the national 100 smart cities programme, it has been appropriated within this programme since they are united by a common vision to use 'urbanisation as a business model' (Borpuzari, 2011: 97) in India. Thus while Dholera presents itself as 'smart', it might become just another ordinary trope in the history of Indian urban planning.

Entrepreneurial urbanisation in Gujarat

For some time now the regional state of Gujarat in India has acted as an 'entrepreneurial state' (Mazzucato, 2013). Through an interlinking of particular types of state policies, stakeholders, institutions and organisations, Gujarat has continually presented itself as open to innovation and investment – as 'India's growth engine and economic powerhouse' and 'the only state in India to emerge as investor friendly even during the world economic downturn' (GIDB, 2014: online). Since 2001 Gujarat has been seen as a state with 'minimum government and maximum governance', which has meant the simplification of bureaucratic processes and the paring down of laws in order to make investment fast and 'friendly'.

This entrepreneurialism stretches across several scales and spaces to make possible the emergence of megacity clusters such as the Delhi–Mumbai Industrial Corridor (DMIC), where Dholera smart city is located. The DMIC seeks to create markets in smart cities through innovation and investment in trade and investment in the region between the two megacities. This 'innovation' was followed through with publicity and marketing by Gujarat state, which launched a biennial trade show, called the Vibrant Gujarat Summit, at which Dholera smart city was first publicly unveiled in 2013. In December 2013, the US-based Smart City Council (which includes companies such as IBM, Microsoft and Cisco as partners) opened its first regional chapter in South India. The purpose was to set a new agenda for smart cities in India and to 'accelerate growth in the smart cities sector by lowering barriers to adoption through thought leadership, outreach, tools and advocacy' (Smart Cities Council, 2013: online).

These strategies of a regional state resonate with the 'distinctive feature' of entrepreneurial cities 'as being proactive in promoting the competitiveness of their respective economic spaces in the face of intensified international (and also, for regions and cities, inter- and intra-regional) competition' (Jessop, 1997: 28). In the case of Gujarat, however, these entrepreneurial strategies are not just at the scale of the city or state, but rather – as noted activist-scholar Arundhati Roy (2012) puts it – as a 'matryoshka doll' that promotes urbanisation, from the city to the nation. Since the May 2014 election results, this is an all-encompassing vision of entrepreneurial 'smart' urbanisation, scaled and duplicated via the 100 smart cities programme. Crucially this entrepreneurial urbanisation while located in the sphere of economic or material gains, is reinforced and perpetuated through a rhetoric of crisis and development. This entrepreneurial urbanisation is both real and imagined through its documents, policies, and masterplans; its grandiose flythrough simulations and glossy promotional material; and through its enforcement over populations and territories.

Dholera, a smart entrepreneurial city?

While entrepreneurialism from the city to the national scale is easily charted into the 100 smart cities programme, this form of entrepreneurial urbanisation also promotes what Goldman calls 'speculative urbanism'. For Goldman (2011: 555), speculative urbanism reflects a 'shift into new forms of "speculative" government, economy, urbanism and citizenship'. In his analysis of the Mysore–Bangalore development corridor, Goldman (2011) notes that the main 'business' of the state is now in creating and extracting surplus value and that it does so by actively engaging in land speculation and active dispossession of those who come in the way of global city projects.

The extraction of surplus value is evident in the pastoral and rural economies and livelihoods. Actually existing Dholera is a small village located in a vast, low-lying ecological area off the Gulf of Khambhat (on the Arabian Sea) in Gujarat (see Figures 4.1 and 4.2). It is one of the 22 villages which will be pooled together to constitute 'Dholera smart city'. This region remains submerged under the sea for part of the year, losing at least 1 cm of its coastline to the sea each day. It is a region of low population density, with 6,532 households and 37,712 inhabitants in the 2001 census. The overall literacy rate in the region is 57 per cent, far lower than the national average of 77 per cent and the Gujarat average of 81 per cent. It is largely inhabited by 'Koli Patels' (at 62 per cent), an indigenous fishing community, and a number of other social groups who have been listed as 'Scheduled and Backward Castes' by the Indian state since 2001 (Senes, 2013). Forty-seven per cent of land in this region is agricultural, with 62 per cent of residents occupied in agriculture. They show a high reliance on subsistence farming and minimum demands for industrial products. Its farmers were promised water from the controversial Narmada Dam built in 2006, but the state's unfulfilled promise has seen increased soil salination and consequently a decline in agricultural productivity over the years. Dholera, then, is a classic *terra nullis* for global investors, politicians and planners since it presents a discourse of 'lack' – of productive agriculture, development and modernity.

Although no ground has been broken for any of the buildings in the master plan, there has been rising speculation on land, with a plethora of advertisements nearby announcing sales of land for private residential purposes or the sale of luxury apartments in new gated communities around Dholera. This is because Dholera smart city will reduce its existing agricultural land from 67 per cent to only 12 per cent and will also be connected by rail to the nearest city, Ahmedabad, 100 km to its north. New proposals put together, for supplying electricity and fresh water by constructing the nearby Kalpasar mega-dam project, supplying industrial trade by the development of a seaport and global business with the construction of an international airport, provide strong indications of a proactive state that is good news for business. Finally, the construction of Dholera is supposed to spearhead economic growth in the region, generating 0.8 million jobs and supporting 2 million inhabitants by the year 2040 (Halcrow Group, personal communication). These promises actively cultivate the national imagination of transforming local economies from their seemingly parochial origins to modern urban identities.

FIGURES 4.1 AND 4.2 The actually existing Dholera region
Source: JAAG (used by permission)

Dholera's speculative urbanism relies upon its simultaneous links to entrepreneurialism and crisis. When Dholera was designed by Halcrow UK, it was labelled and granted planning approval in 2009 as an industrial township in order to promote trade and manufacturing in the region. It was only in December 2012 in a TEDx lecture given by the CEO of the DMIC project, Amitabh Kant, that the idea of 'smart growth' for seven new cities, including Dholera, was presented. He further elaborated,

> To my mind, technology holds the key … digital technology has allowed the world to do urbanisation, and instead of vertical, do horizontal urbanisation. Therefore today's cities not only have to be interconnected, transit oriented, walkable and cycle-able, they have to be the smart cities of the future … It means India can make a quantum leap into the future … it means you can drive urbanisation through the back of your mobile phone.
>
> *(TEDx Talks, 2012: online)*

In this TEDx lecture, then, Dholera achieved a metamorphosis from an 'industrial township' to a 'smart city'. Its smart labelling was made visible thereafter in all the super-simulated promotional videos. In its discursive and material transformations from an industrial to a smart city, through an ambiguous rhetoric of 'mobile phone-driven urbanisation', Dholera was thereafter simultaneously labelled as an industrial city, eco-city and smart city when it was unveiled in the Vibrant Gujarat Summit in January 2013.

'Smart' in the case of Dholera is a highly subjective parameter to be given meaning through a 'global IT powerhouse' (Sharma, 2013), Cisco, which has operationalised the smart city trope through discourses of efficiency, organisation, intelligence and functionality (Hollands 2008). Dholera's 'smart' credentials reflect the fusion of eco-city and networked city ideologies. Its claims to eco-city status include a range of renewable energy initiatives, low carbon footprints, wildlife sanctuaries and so on. Its 'smartness' is presented via features such as 'connected homes', green residential spaces, 'futuristic' malls and marketplaces and an advanced MRT system (ARTIST2WIN, 2013). Its claim to industrial township is vested in the location of a Gujarat Trade Centre in the city and its proximity to the airport, seaports and DMIC. Its claim to be a knowledge city is vested in its entertainment and knowledge zones, university and training centres, super-speciality hospitals and so on. Indeed, Dholera presents such an all-encompassing utopia of a future city that its scaling up to a national level seemed inevitable when it was mentioned in the speech of the Union Budget in February 2013.

> Plans for seven new cities have been finalised and work on two new smart industrial cities at Dholera, Gujarat and Shendra Bidkin, Maharashtra will start during 2013–14. We acknowledge the support of the Government of Japan. In order to dispel any doubt about funding, I wish to make it clear that we shall provide, if required, additional funds during 2013–14 within the share of the Government of India in the overall outlay for the project.
>
> *(IBNLive, 2014: online)*

This announcement, made by the Indian Finance Minister, highlighted Dholera's significance as a new flagship project. Conceived here as a 'smart industrial city' Dholera's identity relies on a 'definitional impreciseness' (Hollands, 2008: 304). 'Industrial' and 'smart' as labels are used interchangeably – the former representing economic reasoning and the latter reflecting globally marketable logics for attracting business and investment.

Dholera's self-congratulatory (Hollands, 2008) rhetoric, evident in its simulations and publicity videos, however, hides the ideological forces and politics behind its making, and the absences and silences that shroud the discourses perpetuated by its most enthusiastic supporters (both public and private sector). Neither the plans nor videos of Dholera, nor the speeches of Narendra Modi, nor the lectures of bureaucrats making presentations refer to actually existing Dholera, which remains as an absent presence, giving the impression of an empty backdrop, a *terra nullis* – the perfect landscape-in-waiting for the smart city.

Speed as a response to crisis

Speed is an essential component of smart urbanism in India. Smart urbanism relies upon the use of speed as an antidote to the rhetoric of crisis. Smart cities are produced through a speeding up of time that condenses the experience of urbanisation to within a few years rather than over a few decades. In other words, smart urbanism in India is not a gradual transformation of social, cultural, material and political landscapes as seen in existing Indian cities; rather it is a material manifestation of speed itself. In this context the 100 smart cities programme in India is a response to a crisis of rapidly increasing urban population, as evident in the smart city concept note:

> The trend of urbanisation that is seen in India over the last few decades will continue for some more time … With an urban population of 31 percent, India is at a point of transition where the pace of urbanisation will speed up. It is for this reason that we need to plan our urban areas well and cannot wait any longer to do so.
>
> *(Government of India, 2014: online)*

Gujarat, however, had begun its path to entrepreneurial urbanisation much earlier than the national smart cities programme through a 'fast tracking' of legal reform. In 2009, the Gujarat government passed a Special Investment Region (SIR) Act, in order to 'fast track' industrialisation of the region. Similar to the legislation enabling a Special Economic Zone (SEZ), the SIR Act (with provisions taken from the Gujarat Town Planning and Urban Development Act, 1973) applies specifically to development within Gujarat on any area of more than 100 km^2 or any Industrial Area with an area of 50–100 km^2. Unlike SEZs, which are largely developed through foreign investment, the Gujarat government has a much larger stake in the SIR, being able to set up government agencies within its area, seen in the designation of the area as an 'industrial township' and controlled by a parastatal agency – the Gujarat Industrial Development Board (GIDB).

In his research on the Bangalore–Mumbai corridor, Goldman notes that the 'historical convergence of neoliberalisation and world-city urbanisation has empowered the new parastatal agencies … to become brokers of large-scale public and private land transfers' (2011: 567). The speed with which the SIR Act was conceptualised and implemented by the GIDB is evident from this timeline – the SIR Act was passed in March 2009, notification of Dholera as an SIR was received in May 2009, masterplans were completed by Halcrow UK in October 2010, a development plan for the SIR was approved in December 2011 and finally land allocation started in December 2012. The new SIR law bypasses India's 1894 Land Acquisition Act, revised to make fair compensation for land acquisition as well as consultation with local self-government institutions mandatory. The SIR Act, however, falls under the Gujarat Town Planning Scheme (GTPS) 1976, which defines land for town planning schemes as deemed to be 'land needed for public purpose' and therefore not subject to mandatory consultation. Land can be acquired by the GIDB under this scheme, and the GIDB need only notify those who are part of Special Investment Regions (SIRs) in order to acquire agricultural land and reallocate it to new urban development.

Dholera thus reflects a radical internalisation of a 'bypass urbanisation' (Bhattacharya and Sanyal, 2011) that circumvents not only the challenges of existing megacities by building new cities, but more crucially also the federal laws (Land Acquisition) of the state. As Watson (2014: 230) argues, 'It is access to land by the urban poor (as well as those on the urban periphery and beyond) that is most directly threatened by all these processes, and access to land in turn determines access to urban services, to livelihoods and to citizenship'. In many ways Dholera outlines the processes through which 'land banks' (Goldman, 2011) are being created across the country to establish the global reach of previously 'parochial' regions. In this the rhetoric of crisis becomes an active discursive tool of the state through which the 'magnitude, speed, and the overarching aura of legitimacy of these new governance endeavours' (Goldman 2011: 575) surface. Indeed these rhetorics become necessary to produce places like Dholera and replicate them elsewhere. It presents a situation that Sassen would call an 'extreme case of key economic operations' (2014: 9) of a neoliberal state, which is playing an ever increasing role in directing and controlling the discourses and practices of urban planning with a proactive role in entrepreneurial urbanisation.

Sluggish delivery

Speed is only half the story of Dholera smart city. As Peck (2002: 348) notes, 'the confident rhetoric of fast-policy solutions and the conviction-speak of neoliberal politicians collide with the prosaic realities of slow (and uncertain) delivery'. Examples of earlier projects in India and elsewhere suggest that most of these city-making projects encounter a number of challenges and bottlenecks with stalled construction and low population that slow down construction. The 'sluggishness' of delivery of these cities is hidden from the discursive, techno-material and social imperatives that are presented to legitimise their speed of delivery. Yet it is the technical and social arenas where they slow down.

Technicalities

Dholera faces several technical roadblocks identified in its impact assessment report. First, the report underlined the high risk of flooding in Dholera, which means that it would cost over Rs 700 crore to do the necessary engineering works for flood mitigation (Senes, 2013). Second, Dholera SIR will be built close to a blackbuck habitat and would therefore lead to irreversible loss of biodiversity. These challenges provided grounds for the withdrawal of several early investors from Dholera. The flood assessment report also led to the abandonment of plans to build an international airport in Dholera. Plans for the international seaport were also abandoned, as well as the nearby Kalpasar dam project. The Indian state has instead proposed to raise the height of the existing Narmada dam project to provide water to the region. This in turn has spurred renewed national movements among activists in support of farmers and indigenous populations who will be disposed far away from Dholera in order to supply water to the region.

Land as the final frontier

The key stumbling-block in Dholera's smart urbanism however has been the question of land. This is recognised in India's recent smart city concept note which states that 'land, laws need to be liberal in India' (Government of India, 2014). However, agricultural land particularly in regions of declining agricultural productivity with lower population density (and hence presumed as low potential for local resistance), makes land acquisition relatively straightforward. While in the Bangalore–Mysore corridor, Goldman argues that 'peasants [are] the final frontier in city-making' (Goldman, 2011), the production of Dholera smart city makes land the final frontier of urbanisation. Dholera's mechanism of land acquisition, imposed by a rule of law vested in the SIR Act, becomes a state-orchestrated exercise of sovereign rule. Levien (2013) describes this process as a 'regime of dispossession', where socially and historically specific constellations of state structures of bureaucracy and governance produce particular patterns of dispossession of peasants and landless farmers. In Dholera, these initiate a new 'regime of entrepreneurial urbanisation' whereby land is acquired for a 'public good' and 'delinked from capitalist production, by making it available for capitalist space of any kind' (Levien, 2013: 199). This model of extracting surplus value from land by converting the rural/agricultural commons to state property to then sell as it wishes to the private sector 'aggressively uses the rules of eminent domain to acquire land from the few who own land and the many who

FIGURE 4.3 JAAG methods of social action
Source: JAAG (used by permission)

thrive off the land, and place them on the new multi-lane highway to elsewhere' (Goldman, 2011: 577). However, written within this system is also its own unfolding, where the monetisation of land values must continuously confront new struggles of citizenship and rights to this land.

JAAG Land Rights Movement

In Gujarat, farmers' cooperatives have recently begun to organise under a coalition called *Jameen Adhikar Andolan Gujarat* (JAAG) or Gujarat Land Rights Movement to claim their rights to the commons – agricultural land, common property, fishing areas and pasture land, among many others. JAAG methods of social action have included public protests, marches, putting up notices outside the villages barring state officials from entering their land and several other peaceful demonstrations (Figure 4.3). JAAG achieved some success in a neighbouring region when 44 villages therein were notified as an 'auto and knowledge hub' under the SIR Act (ET Bureau, 2013). In 2013, as a result of JAAG protests, the Gujarat government was forced to withdraw 36 of the total 44 villages from the SIR notification (Counterview, 2013). In April 2014, another nearby SIR was withdrawn after several protests. This gave hope to the JAAG campaigning against the Dholera SIR notification, resisting the release of their land holdings to the Gujarat state government without any real compensation.

While the state is increasingly asserting its sovereignty through a rule of law and has criminalised several aspects of JAAG's social action and protest, many of JAAG's efforts have been to slow down the process of building Dholera by direct participation in state-prescribed processes of bureaucracy. On their part, JAAG has provided challenges to the instrumental and technocratic tools embodied in the Environmental Impact Assessments (EIAs) notifications that require mandatory public consultations for large-scale township projects. The EIA public hearing for Dholera, in January 2014, was heavily policed and video-taped as per the provisions of the EIA notification. Over 500 members of the public, which included JAAG activists, farmers and several other members, exercised their right to democratic participation by attending the public hearing and raising their objections about the project with the Gujarat state authorities.

JAAG noted that the draft EIA prepared by Senes (2013), the state-appointed EIA consultants, had several instances of misinformation and misrepresentation of facts associated with the used of outdated maps. This made it more of a 'bureaucratic arrangement' (Narain, 2009) rather than a participatory tool through which states could listen to local communities and respond to environmental contingencies. Where new compensatory land is provided, farmers claim that most of this is infertile or disconnected from irrigation canals that are essential for agriculture, and that it would take years of work to make these cultivable. Indeed, in several cases, the compensatory land allocated to farmers was based on 100-year-old maps and has already been claimed by the sea (JAAG, personal communication). JAAG activists argue that Dholera SIR will lead to large-scale transformations in livelihoods of farmers, benefiting those with larger parcels of land and dispossessing small-scale subsistence farmers and the landless. Indeed, while these important issues were raised in the public hearing, their ultimate disregarding in the final approval for Dholera highlighted how farmers and indigenous populations have now become the 'weakest links' in smart city making.

Gujarat police in retaliation have issued warrants against several activists and arrested them (Telegraph India, 2013), denied them licence to stage peaceful protests and engaged in several instances of harassment and bullying with farmers and activists. More recently a leaked Indian Intelligence Bureau report named JAAG as one of the organisations 'under watch' for engaging in 'anti-development activities' (Pathak, 2014). The state has also begun to issue notices to several farmers to either hand over the land and take whatever compensatory land is offered or be prepared to be evicted by the state officials. Violently imposed upon landscapes and populations who were presented as 'lacking' in development and therefore ideal for a 'makeover', smart city Dholera has turned into a 'temporary "state of exception", with both its attendant suspensions of civil and human rights as well as their institutionalisation into government practices' (Goldman, 2011: 555).

Afterword: is Dholera a prototype of India's smart urbanism?

Dholera is an indication of the future of India's urbanisation. It illustrates how an entrepreneurial state attempts to proactively attract capital and spur growth through a smart urbanism trope. This trope is as much textual (policy documents) and verbal (political speeches and interviews) as it is representational (promotional brochures and videos of smart cities). By looking at how this trope emerges from below the scale of the nation to stand for a model of entrepreneurial urbanisation nationally, I have suggested that Dholera smart city shows how 'nation states have the capacity to enforce their truth games' (Chakrabarty, 1992: 338) via 'self-justificatory narratives' of modernity and development.

Following Bunnell's (2002) observations on Malaysia's 'intelligent cities', the 'broad ideological underpinning of strategies to realise smart urbanism in India – liberalisation and modernisation – show similar continuity' in Dholera. As in the case of 'African urban fantasies' (Watson, 2014), the assumption in Dholera is that these new cities are built on 'empty land', thereby evading public and democratic debate on mass-scale expulsions of marginalised citizens from their land and livelihoods. These politics place regional states such as Gujarat at the nexus of dispossession, modernisation and liberalisation in India.

My conversations with JAAG members in December 2013 suggested that Dholera had never held any local associations with smart urbanism. Even as Dholera was being mentioned in the 2013 Union Budget, unveiled in the Vibrant Gujarat Summit and globally circulated via striking flythroughs as a smart city, farmers in JAAG and local activists continually referred to it as Dholera SIR – suggesting that it was the '(mis)rule of law' (Holston, 2008) vested in the SIR Act and not smart urbanism that was the everyday context of local struggles. Now, as farmers continue to stake their claims to their homes and livelihoods in the region, they have raised judicial challenges around the constitutional legitimacy of the SIR Act in bypassing the federal Land Acquisition Act. These struggles show how top-down visions of smart urbanism produce new forms of dispossession and citizenship struggles around law and legality. Simultaneously it also shows that while the trope of the smart city might be new in India, the struggle for citizenship rights and social justice at the grassroots is certainly not a new one.

Dholera is yet to be built, but its discursive and material construction as a smart city provides us an insight into the future of 100 new smart cities proposed by the Indian government. In many ways, Dholera is distinct from the 100 smart cities programme. It predates the initiative and the national election. It is one of the few smart cities which has financial backing from the Union, while the rest of the 100 smart cities will receive only £7million each from the state and will therefore compete in the global economy to find investors. But as an experiment in smart urbanism, Dholera still remains key to the 100 smart cities mission and has shaped it in three distinct ways.

First, the enormous costs and challenges of finding investors for Dholera might have contributed to paring down the national programme on smart cities. The recent concept note from the Ministry of Urban Development (Government of India, 2014) proposes that most of the 100 smart cities will be made by retrofitting a range of small and medium-sized cities with smart transport, housing and infrastructure – not by building new cities from scratch. The scaling down of the national programme to include retrofitted cities of different sizes suggests a scaling down of the entrepreneurial smart city from the new and grandiose urban planning experiments proposed in Dholera, Shendra-Bidkin and other new cities along the DMIC. Indeed several smart cities claiming to be 'new' are those now rebranding themselves from earlier labels of eco-city (Lavasa), finance city (GIFT) and new townships (Rajarhat).

Second, Dholera illustrates how a regional state's innovation in using a rule of law can spur entrepreneurial urbanisation at a national scale. Several other regional states such as Rajasthan, Maharashtra and Odisha are now in the process of putting in place their own versions of the SIR Act to become entrepreneurial states. Indeed since JAAG has challenged the SIR Act in court as unconstitutional and a violation of the federal Land Acquisition Act, the Indian parliament is now deliberating a change to its Land Acquisition Act to remove the clauses on consent and compensation inserted by the previous government. The Indian government is now also bringing in changes in the Foreign Direct Investment (FDI) law to allow direct investment in construction. This means enforcing through a rule of law, a self-sustaining myth of urbanisation as a 'good business model', which increasingly represses the articulation of resistance and social action among marginalised groups.

Finally, resistance action around Dholera often has little to do with its smart credentials, but rather with more traditional struggles around regimes of accumulation and dispossession related to land. Dholera thus is not the only smart city facing local resistance. Recently the media reported on students resisting the construction of a new smart city, Haollenphai, that sought to acquire Manipuri tribal land in north-eastern India. Social action and resistances around industrial townships, SIRs, knowledge cities and other utopian planning tropes are making translocal connections of solidarity to challenge the state via social and legal mechanisms. Shendra-Bidkin and Manesar-Dhow – two other proposed smart cities – have also seen rising social action among farmers, rarely reported in the media. Merging under the umbrella of the National Alliance for People's Movements (NAPM), these movements against the smart city are part of a bigger movement to resist the increasing neoliberalisation of land and the commons. Land and its acquisition are now more strongly linked than ever before with questions of citizenship, inclusion and belonging to the wider Indian polity.

Dholera might well be part of 100 top-down utopian visions of entrepreneurial urbanisation, but it has also inadvertently produced innovation and enterprise at the grassroots. In that context Dholera's entrepreneurialism predates the smart city trope. Reflection, learning and innovation through knowledge and awareness about law and its practices, about bureaucratic processes and democratic encounters with the state are already familiar to the farmers in Dholera. In those terms farmers in Dholera and other proposed smart cities are innovators of change, since they must continuously find ways to reach out locally, nationally and globally in order to gain a collective voice for their struggles. As one of the JAAG activists said to me, 'I am glad you are writing about this, since by doing so you are helping our cause.' It seems, however, that we still know little about grassroots forms of transformative learning, knowledge and action that can challenge the normative production of smart urbanism in India. Here, the available tools of analysis using political economy, policy mobility and postcolonial urbanism need to be complemented by ethnographic details on the everyday struggles faced by those at risk of being excluded from India's urbanisation. It means examining how the 'population' referred to in the official reports become 'citizens', claiming their rights to livelihoods and landscapes as they encounter the smart city. It means examining how the smart city will be built not by digital citizenships, but by 'insurgent citizens' (Holston, 2008) living on its margins – socially, geographically, legally and economically. Crucially, it means understanding how a right to the city is inherently connected to a right to commons as political and social action gather momentum against the smart city.

Acknowledgements

The author would like to express her heartfelt gratitude to JAAG members for their kindness in answering questions via phone and email conversations, discussing the issues around land acquisitions for Dholera and for providing information about the various aspects of their movement. The author is also grateful to JAAG for supplying photos for inclusion in this chapter.

References

3iNetwork (2009) *Land: A Critical Resource for Infrastructure. India Infrastructure Report*. New Delhi: Oxford University Press. Available at: www.iitk.ac.in/3inetwork/html/reports/IIR2009/IIR_2009_Final_July%2009.pdf [Accessed 1 July 2013].

ARTIST2WIN (2013) Dholera SIR – Future Smart City [online video], 11 June. Available at: www.youtube.com/watch?v=jOFpWFLSqgU [Accessed 19 December 2013].

Bhattacharya, R. and Sanyal, K. (2011) Bypassing the squalor: new towns, immaterial labour and exclusion in post-colonial urbanisation. *Economic and Political Weekly* (July 30): 41–48.

Borpuzari, P. (2011) The need of the hour is to create integrated cities: interview with Amitabh Kant [online]. *Entrepreneur* 3 (4): 95–97. Available at: http://entrepreneurindia.in/runandgrow/strategy-runandgrow/the-need-of-the-hour-is-to-create-integrated-cities/10598/ [Accessed 1 May 2014].

Bunnell, T. (2002) Multimedia utopia? A geographical critique of high-tech development in Malaysia's multimedia super corridor. *Antipode* 34 (2): 265–295.

Chakrabarty, D. (1992) Provincializing Europe: postcoloniality and the critique of history. *Cultural Studies* 6 (3): 337–357.

Chatterjee, P. (2004) *Politics of the Governed: Reflections on Popular Politics in Most of the World*. New York: Columbia University Press.

Choe, K., Laquian, A. and Kim, L. (2008) *Urban Development Experience and Visions: India and the People's Republic of China*. ADB Urban Development Series. Manila: Asian Development Bank. Available at: http://citiesalliance.org/sites/citiesalliance.org/files/ADB_Urban-Visions.pdf [Accessed 13 October 2013].

Choplin, A. and Franck, A. (2010) A glimpse of Dubai in Khartoum and Nouakchott: prestige urban projects on the margins of the Arab world. *Built Environment* 36 (2): 192–205.

Comaroff, J. and Comaroff, J. (2006) *Law and Disorder in the Postcolony*. Chicago and London: University of Chicago Press.

Counterview (2013) Opposition to land acquisition in Gujarat picks up, with JAAG expanding its wings towards Dholera SIR [online]. Available at: www.counterview.net/2013/12/opposition-to-land-acquisition-in.html [Accessed 20 December 2013].

Datta, A. (2012) India's eco-city? Urbanisation, environment and mobility in the making of Lavasa. *Environment and Planning C* 30 (6): 982–996.

DMIC (2008) Delhi–Mumbai Industrial Corridor [online]. Available at: www.dmic.co.in [Accessed July 2013].

ET Bureau (2013) Farmers protest against investment regions in Gujarat [online]. Available at: http://economictimes.indiatimes.com/news/politics-and-nation/farmers-protest-against-investment-regions-in-gujarat/articleshow/27658543.cms [Accessed 21 December 2013].

Fishman, R. (1982) *Urban Utopias in the Twentieth Century: Ebenezer Howard, Frank Lloyd Wright, and Le Corbusier*. Cambridge, MA: MIT Press.

Freestone, R. (ed.) (2000) *Urban Planning in a Changing World: The Twentieth Century Experience*. New York: Taylor & Francis.

GIDB (2014) Gujarat Industrial Development Board [online]. Available at: http://www.gujaratindia.com/business/investment-destination.htm [Accessed 3 May 2014].

Goldman, M. (2011) Speculative urbanism and the making of the next world city. *International Journal of Urban and Regional Research* 35 (3): 555–581.

Government of India (2014) Draft concept note on smart city scheme. Delhi: Ministry of Urban Development.

Graham, S. (2000) Constructing premium networked spaces: reflections on infrastructure network and contemporary urban development. *International Journal for Urban and Regional Research* 24 (1): 183–200.

Hall, T. and Hubbard, P. (1996) The entrepreneurial city: new urban politics, new urban geographies? *Progress in Human Geography* 20 (2): 153–174.

Hollands, R. (2008) Will the real smart city please stand up? Intelligent, progressive or entrepreneurial? *City* 12 (3): 303–320.

Holston, J. (1989) *The Modernist City: An Anthropological Critique of Brasilia*. Chicago: University of Chicago Press.

Holston, J. (2008) *Insurgent Citizenship: Disjunctions of Democracy and Modernity in Brazil*. Princeton: Princeton University Press.

IBM (2010) Smarter cities for smart growth [online]. Available at: www-935.ibm.com/services/us/gbs/bus/html/smarter-cities.html [Accessed 2 December 2013].

IBNLive (2014) Full text Union Budget 2013–14: read Finance Minister P. Chidambaram's budget speech [online]. Available at: http://ibnlive.in.com/news/full-text-union-budget-201314-read-finance-minister-p-chidambarams-budget-speech/375639-7-255.html [Accessed: 1 May 2014].

Infotech (2014) ICT accounts for less than 1% of civic budget in proposed smart cities in India [online]. Available at: www.infotechlead.com/networking/ict-accounts-less-1-civic-budget-proposed-smart-cities-india-26830 [Accessed 5 January 2015].

Jessop, B. (1997) The entrepreneurial city: re-imaging localities, redesigning economic governance, or restructuring capital? In N. Jewson and S. MacGregor (eds) *Realising Cities: New Spatial Divisions and Social Transformation*. London: Routledge, pp. 28–41.

Jessop, B. (2008) The enterprise of narrative and the narrative of enterprise: place marketing and the entrepreneurial city. In T. Hall and P. Hubbard (eds) *The Entrepreneurial City*. Chichester: Wiley, pp. 77–99.

Jessop, B. and Sum, N. (2000) An entrepreneurial city in action: Hong Kong's emerging strategies in and for (inter)urban competition. *Urban Studies* 37 (12): 2287–2313.

Kant, A. (2013) In India, leaders are building smarter cities from the ground up [online], 19 September. Available at: http://asmarterplanet.com/blog/2013/09/in-india-leaders-are-building-smarter-cities-from-the-ground-up.html [Accessed 2 December 2013].

Kargon, R. and Mollela, A. (2008) *Invented Edens: Techno-Cities of the 20th Century*. Cambridge, MA: MIT Press.

Kitchin, R. (2013) The real time city: big data and smart urbanism. *GeoJournal*. (doi: 10.1007/s10708-013-9516-8).

Kundu, A. (2014) India's sluggish urbanization and its exclusionary development. In G. McGranahan and G. Martine (eds) *Urban Growth in Emerging Economies: Lessons from the BRICS*. London: Routledge, pp. 191–231.

Kundu, R. (forthcoming) Inserting local agendas in the planning of new town Rajarhat, Kolkata. In Ayona Datta and Abdul Shaban (eds) *Mega-urbanization in the Global South: Fast Cities and New Urban Utopias of the Postcolonial State*. London: Routledge.

Levien, M. (2013) Regimes of dispossession: from steel towns to Special Economic Zones. *Development and Change* 44 (2): 381–407.

Liscombe, R. (2007) In-dependence: Otto Koenigsberger and modernist urban resettlement in India. *Planning Perspectives* 21: 57–178

Luque-Ayala, A. and Marvin, S. (2015) Developing a critical understanding of smart urbanism? *Urban Studies* (doi: 10.1177/0042098015577319).

Macfarlane, C. (2011) *Learning the City: Knowledge and Translocal Assemblage*. Oxford: Wiley-Blackwell.

McKinsey Global Institute (2010) *India's Urban Awakening: Building Inclusive Cities, Sustaining Economic Growth*. New Delhi: McKinsey Global Institute.

McKinsey Global Institute (2011) *Big Data: The Next Frontier for Innovation, Competition, and Productivity*. New Delhi: McKinsey Global Institute.

Maeng, D. M. and Nedovic-Budic, Z. (2008) Urban form and planning in the information age: lessons from literature. *Spatium* 17–18: 1–12.

Mazzucato, M. (2013) *The Entrepreneurial State: Debunking Public vs. Private Sector Myths*. London: Anthem Press.

Menon, A. (2013) Integrating Dholera SIR through ICT. In 'Vibrant Gujarat Summit', seminar on 'Developing Integrated, Smart and Sustainable Cities, with a Focus on DMIC and SIRs: Seminar proceedings and way forward, 11–13 January' [online]. Available at: www.gidb.org/downloads/smart_cities_seminar_proceedings_report.pdf. [Accessed 30 May 2014].

Narain, V. (2009) Growing city, shrinking hinterland: land acquisition, transition and conflict in peri-urban Gurgaon, India. *Environment and Urbanization* 21 (2): 501–512.

Olds, K. (2001) *Globalization and Urban Change: Capital, Culture, and Pacific Rim Mega-Projects*. Oxford: Oxford University Press.

Panchal, P. (2014) Dholera SIR 2 times bigger than Delhi [online video]. Available at: http://youtu.be/qfzBYSIc37w [Accessed 6 June 2014].

Pathak, M. (2014) The Narendra Modi model of development [online]. Available at: www.livemint.com/Politics/jidgEODmTuivukTLuJHSTN/The-Narendra-Modi-model-of-development.html [Accessed 30 May 2014].

Peck, J. (2002) Political economies of scale: fast policy, interscalar relations, and neoliberal workfare. *Economic Geography* 78 (3): 331–360.

Peck, J. (2005) Struggling with the creative class. *International Journal of Urban and Regional Research* 29 (4): 740–770.

Rapoport, E. (2014) Globalising sustainable urbanism: the role of international masterplanners. *Area* (doi: 10.1111/area.12079).

Roy, A. (2011) The blockade of the world-class city: dialectical images of Indian urbanism. In A. Roy and A. Ong (eds) *Worlding Cities: Asian Experiments and the Art of Being Global*. Oxford: Wiley-Blackwell, pp. 259–278.

Roy, A. (2012) Capitalism: a ghost story [online]. Available at: www.outlookindia.com/article.aspx?280234 [Accessed 2 May 2014].

Roy, A. and Ong, A. (eds) (2011) *Worlding Cities: Asian Experiments and the Art of Being Global*. Oxford: Wiley-Blackwell.

Sarkar, T. and Chowdhury, S. (2009) The meaning of Nandigram: corporate land invasion, people's power, and the left in India. *Focaal* 54: 73–88.

Sassen, S. (2014) *Expulsions: Brutality and Complexity in the Global Economy*. Cambridge, MA: Harvard University Press.

Senes Consultants (2013) DRAFT EIA of Dholera Special Investment Region (DSIR) in Gujarat [online]. Available at: www.gpcb.gov.in/pdf/DMICDC_DHOLERA_SPECIAL_INVEST_DSIR_EIA.PDF [Accessed 6 December 2013].

Sharma, A. K. (2013) Welcome address. In 'Vibrant Gujarat Summit', seminar on 'Developing Integrated, Smart and Sustainable Cities, with a Focus on DMIC and SIRs: Seminar proceedings and way forward, 11–13 January' [online]. Available at: www.gidb.org/downloads/smart_cities_seminar_proceedings_report.pdf [Accessed: 30 May 2014].

Shatkin, G. (2007) Global cities of the South: emerging perspectives on growth and inequality. *Cities* 24 (1): 1–15

Shaw, A. (2009) Town planning in postcolonial India, 1947–1965: Chandigarh re-examined. *Urban Geography* 30 (8): 857–878

Sheppard, E., Leitner, H. and Maringanti, A. (2013) Provincializing global urbanism: a manifesto. *Urban Geography* 34 (7): 893–900.

Smart Cities Council (2013) Welcome to the Smart Cities Council – India [online]. Available at: http://smartcitiescouncil.com/india [Accessed 1 May 2014].

TEDx Talks (2012) Amitabh Kant at TEDx Delhi [online video], 17 December. Available at: www.youtube.com/watch?v=8BvMybtJ1-E [Accessed 2 February 2014].

Telegraph India (2013) Gujarat crackdown on land protest [online]. Available at: www.telegraphindia.com/1130819/jsp/nation/story_17246207.jsp#.U5B03ygmaKI [Accessed 1 May 2014].

Townsend, A. (2013) *Smart Cities: Big Data, Civic Hackers, and the Quest for a New Utopia*. New York: W.W. Norton.

Vanolo, A. (2014) Smartmentality: the smart city as disciplinary strategy. *Urban Studies* 51 (5): 883–898.

Watson, V. (2014) African urban fantasies: dreams or nightmares? *Environment and Urbanization* (26): 215–231.

5
GETTING SMART ABOUT SMART CITIES IN CAPE TOWN
Beyond the rhetoric

Nancy Odendaal

South African municipalities have struggled implementing what has been perceived as the golden key to effective service delivery and governance: smart, connected cities. Early on, the motivation for incorporating e-governance and digital infrastructure development into urban objectives was tied to city marketing as well as broader strategic objectives, which in the South African context entail poverty alleviation and digital democracy. This resonates with other global South contexts. These initiatives were also circumstantial. Smart city strategies in Cape Town and Durban, for example, coincided with local government restructuring that resulted in both cities becoming metropolitan governance entities and larger geographically. Given the South African constitutional focus on developmental local government, electronic governance strategies were orientated towards more inclusionary forms of development.

The recent re-emergence of the smart city discourse, in South Africa and elsewhere, is associated with high-profile interventions by multinational companies such as IBM and Cisco, and provides an opportunity to revisit this discourse. This chapter explores whether the surge of private sector interest has impacted on smart city objectives in Cape Town. In doing so, it explores the textures of 'smart' that represent a more grounded perspective on the opportunities and limitations of digital governance in situations of profound inequality. This exploratory study shows that day-to-day operations and manifestations are more coarsely grained than the media discourse suggests, in many ways mirroring the 'real' city.

Introduction

In 2002 the City of Cape Town undertook a Digital Divide Assessment entitled 'Taking Stock and Looking Ahead'. The research was done by bridges.org, a Non-Governmental Organisation (NGO) known then for pursuing digital inclusion internationally. The survey involved 2,000 people from civil society, business, academic institutions and government agencies. The aim was to understand qualitative and quantitative constraints to technology access. It had four main objectives: gauge real access to ICT, assess the needs of Cape Town's people and organisations for future ICT services, identify opportunities to improve access and identify constraints (Bridges.org, 2002). Interviews with staff from the City of Cape Town and bridges.org confirmed that the study was to feed into the smart city strategy, released in 2001, aimed at achieving broader developmental goals.

More than ten years later, the City of Cape Town, together with the Swedish consul in Cape Town, hosted a smart city summit in 2012. The opening address by the Swedish ambassador emphasised the function of technology as an enabler of transparency and fighter of corruption. The emphasis was on city governance and climate change, featured boldly as part of a normative agenda that would need to be addressed through improved ICT-enabled city administration.

The two initiatives share the same terminology but differ in objectives and perspectives. In the interim the city had also embarked on a universal access strategy with the Western Cape provincial government and, perhaps more poignantly, technology had evolved dramatically, especially with regards to ubiquitous access to smart phones.

The notion of the smart city is enjoying resurgence since the early explorations of digital cities. It is a slippery concept, open to abuse and varied translations (Hollands, 2008). There is surprisingly little theoretical work on the notion of smart. As a concept it pops up in literature on cities generally; the e-governance work was quite dominant in the 1990s and recent work suggests a concern with resultant increased surveillance and the impact that has on urbanism (Graham, 2010; Firmino and Duarte, 2010). This chapter examines the evolution of the smart city discourse in South Africa by focusing on Cape Town. There are three distinctive contextual elements worth highlighting here: a rights-based Constitution that advocates access to information and freedom of expression, decentralised local government with a firm developmental mandate and, finally, a messy and unresolved telecommunications policy environment that has had profound impacts on digital access. These are important inputs into the aim of this chapter, which is to understand the landscape of smart city initiatives within the Cape Town metropolitan area and consider future implications for inclusive governance.

The method employed here is to examine current smart initiatives in Cape Town in relation to broader discourse trends and relate these back to the South African policy mandate of enabling more inclusive cities. Empirically it takes a broad-brush approach in revealing breadth, rather than depth, given that the aim is to understand typologies of smart city interpretations as they manifest in the 'real' city, i.e. what does the smart city *really* look like? It draws on official documentation, published research and interviews with key stakeholders from the City of Cape Town. The intention is to reveal whether manifestations of the smart city discourse actually exist. The unravelling of the many textures of smart Cape Town is also intended to reveal some of the tensions between discourse and practice. The first part provides a frame for these narratives internationally, with some emphasis on the global South.

The smart city

> 'I come from I.B.M. research. I'm attracted to large, complex systems,' Mr. Banavar said. 'Can you think of a system that is more complex than a city?'
> *(Singer, 2012: online, writing for the* New York Times*)*

Early literature on smart cities in developing counties emphasised the e-governance dimension, and was closely related to the push for information and communication technologies for development (or ICT4D). E-government options include using web portals for dissemination of information and creating options for input into policy documents. Using technology to enable more efficient interdepartmental coordination, better communication and more effective billing systems remains a core part of the smart agenda, which has extended to infrastructure management. Providing and sharing information electronically and enabling municipal transactions online are considered essential to contemporary urban life (Odendaal, 2003; Aurigi, 2005). Measuring and encouraging the presence of the information and communication technology (ICT) industry, attracting such skills and incubating associated enterprise in the creative sector have been more substantive outcomes (Odendaal, 2003). In Africa, e-governance had become closely aligned with decentralisation agendas as bilateral agencies saw potential for greater local government autonomy in the absence of central state support (Misuraca, 2007).

Early practical limitations to achieving e-governance in the South in particular relate to the digital divide and obstinate organisational cultures that resist digitisation. In South Africa, the City of Cape Town's Digital Divide Assessment (Bridges.org, 2002) identified a number of constraints to achieving 'smart city' status, ranging from physical access to technology affordability, the availability of appropriate technology, capacity and training to socio-cultural factors such as ease with computing, age, gender dynamics and general cultural exposure. The 'digital divide' issue has dissipated somewhat with ubiquitous mobile phones.

Watson (2014) focuses attention on the discourse of 'smart' as employed by multinational consultancy firms used to promote new, often edge, developments in African cities. Here the notion of smart, together with other fashionable terms such as 'eco' and 'world-class', are part of a marketing language intended to hold sway with politicians and investors, devoid of any references to the actual contexts within which these spaces are created. In India, the Modi regime plans on creating 100 smart cities in India in ten years; these are defined as 'cities that leverage data gathered from smart sensors through a smart grid to create a city that is liveable, workable and sustainable' (Business Standard, 2014). Private companies such as IBM and Cisco are touted to invest in smart grid infrastructure while the government of Singapore is claimed to be interested in supporting the construction of ten smart cities on the Delhi–Mumbai industrial corridor. The *Business Standard*'s (2014) Sunil Seth speaks of it as a 'fuzzy, New Millennium fantasy', with existing technology hubs such as Bangalore and Hyderabad's technology districts coexisting with slums and *chawls*.

References to social development are tenuous and vaguely technically determinist. IBM's Smarter Cities programme makes three connections that apply to the global South: a celebration of how the last 20 years' technology has enabled improved quality of life; an emphasis on climate change and the fact that vulnerability to floods and storms in over half of the developing world is a reality; and the need for improved governance. Technology is the answer. According to IBM it enables integration of services, disaster management and joined-up governance. Importantly, it increases competitiveness through improved quality of life and a vibrant economic climate (IBM, 2014).

The addition of natural resource management in this checklist of what many would consider good urban management is a small departure in smart city history. On the one hand it speaks directly to the *other* smart city, the smart growth imperative, whilst also reintroducing a normative element into digital governance. One of IBM's most visible success stories in this regard is the establishment of a central operations centre in Rio de Janeiro, Brazil, following a flash flood in 2007 (IBM, 2014). Critique of the Rio initiative considers it within the context of preparations and contestations around the 2014 Fifa World Cup and the Olympic Games in 2016, seeing it as veiled reassurance to investors and Fifa and Olympic officials. As in most socio-economically divided cities, the concern is that only well-off neighbourhoods benefit and that it is an interim measure that detracts from real infrastructural problems. A concern with privacy and surveillance has also been expressed (Singer, 2012).

The Rio control centre is an important milestone in IBM's expansion into the local government market. As part of its strategy to raise its annual revenue to more than $150 billion, its Smarter Planet Unit (which includes the Smarter Cities Business) is projected to generate an income of $10 billion by 2015 (Singer, 2012). Rapid urbanisation in the global South is seen as an opportunity for technology multinationals, with an anticipated market value expected to be $1.5 trillion by 2020 (Frost and Sullivan, 2014).

Three trends in contemporary discussions on smart cities in the media and literature are discernible, and are used as a thematic frame to reflect on Cape Town in this chapter. The first is a broadening that seeks a stronger engagement with the social and cultural coordinates of urbanity. In many cases this is marketing language used to augment corporate agendas, a visual language that emphasises global connection. Whether digital technologies enable inclusion and empowerment would be important questions in this regard.

The second is an engagement with natural sustainability and specifically climate change, reinforcing the relationship between livelihoods, disaster management and digital monitoring. Threats to livelihoods have in many cases necessitated social mobilisation, often using communication technologies. Thus, two angles are therefore important: do policy initiatives include digital tools to mitigate and monitor, and are climate change impacts mediated from the bottom up using ICTs?

The third shift is an explicit acknowledgement of other infrastructure services. The relationship between ICT and other utilities has always been implicit. Municipal utility billing systems rely heavily on centralised information capture and processing, for example. Explicit reference to technology-enabled management of services and digital innovations such as smart grids focus on the interrelationship between utility parts. A third question, then, is whether converging technology and the increasing interests of multinationals could lead to increased centralised governance.

Smart Cape Town

Socio-economic context

Cape Town is the second largest city in South Africa (after Johannesburg), with a population of just over 3.7 million (StatsSA, 2011). Its regional GDP per capita compares to that of Mexico City and Naples, with a value of $15,250 per capita, 40 per cent more than the national average (OECD, 2008). Outside the Gauteng City Region (which includes Johannesburg and Pretoria), it is the only major city-region that has increased its share of national output since the demise of apartheid. Its economic base has shifted to that of a service-based economy. The tourism and hospitality industry is a big part of this growth, while creative and knowledge-intensive industries have contributed more substantially recently (OECD, 2008).

Despite these positive macro-indicators, the city suffers from serious inequalities. Only 76 per cent of the labour force is employed, with just over a third (35.7 per cent) of households earning an income of less than $ 300 per month (City of Cape Town, 2012b). The growth in the service sector does not necessarily benefit the low-and medium-skilled population living on the outskirts of the city.

The city's population has increased by almost 30 per cent since 2001, with a large number of in-migrants from other parts of the country living in informal settlements on the city's south-eastern fringes. Fourteen per cent of households live in informal dwellings; there has been a large growth in the number of households living in informal dwellings in the backyards of formal homes also. Seventy-eight per cent of the city's population is formally housed. With informal living conditions come disparities in service infrastructure access: 3.7 per cent of households do not have access to electricity for lighting; 8.8 per cent have no access to sanitation on site; and there are 232,027 households registered as indigent (City of Cape Town, 2012b). Service access is highly differentiated spatially. In Khayalitcha, home to 391,749 people in the south-east of the city, only 45 per cent of households live in formal dwellings, 62 per cent of households have access to piped water in their dwelling or inside their yard and 72 per cent of households have access to a flush toilet connected to the public sewer system (City of Cape Town, 2012b).

The sanitation situation has been one of the most controversial service delivery issues in recent years. The number of households using bucket toilets increased from 34,200 in 2001 to 48,500 in 2011 (City of Cape Town, 2012b). Inadequate maintenance of communal toilets, the local government's supply of temporary rather than permanent facilities and continued implementation of the bucket system has sparked what are now termed 'poo protests' in the city (Cape Times, 2014), with individuals flinging human faeces collected in bucket toilets on to one of the national freeways into the city (Cape Times, 2013a) and depositing human waste in the foyer of its international airport (Cape Times, 2013b). Other efforts at activism reflect the more systematic approach of the Social Justice Coalition (SJC), explored below.

The telecommunications profile of the city shows high access to mobile telephony (93 per cent of the population has access to a cell phone), but very low computer access (37.9 per cent) and an even lower rate of access to telephone land lines (34 per cent). Just over 50 per cent of the city's population has no access to the Internet.

Provincial ICT strategies

In the Western Cape, the twin-pillared approach to economic development detailed in the province's Strategic Plan (WCPG Office of the Premier, 2010) entails creating an enabling environment for business and supporting specific growth sectors. Growth of the knowledge industry, a high concentration of educational institutions as well as growth of the service sector contribute to a favourable environment for online enterprise. The Western Cape Economic Development Partnership, a non-profit entity that comprises organisations from business, industry, civil society and academia, has formulated a long-term vision that focuses on the province as a centre of innovation and creativity in ICT programming. Its 'One Cape 2040' vision stipulates that provincial government is best placed to enable the provision of relevant and cost-effective ICT services to the citizens through household connections ('last mile'), ICT access in public facilities and through community sectors. The role of local government, according to the plan, is to facilitate access to ICT and related services via integrated service nodes and ICT access at public facilities and buildings as well as hotspots in public places (Western Cape Economic Development Partnership, 2013).

The vision of a 'Connected Cape' is developed in the Western Cape Department of Economic Development and Tourism's Broadband Implementation Plan. The plan focuses on connected citizens, connected government and a connected economy through the development of infrastructure, ensuring readiness to use this infrastructure and driving usage of these services. The role of government is therefore seen as paramount to facilitating and enabling the development and expansion of ICT infrastructure.

Bulk infrastructure roll-out is facilitated through the City of Cape Town and Western Cape provincial government partnering in providing broadband connectivity between provincial and local government buildings as well as schools in some of the city's most disadvantaged areas. At the conclusion of this project (planned over ten years) high-speed Internet connectivity will then be available to a total of 130 city buildings (including 25 clinics) and 45 Western Cape government buildings (Western Cape Government, 2013).

This backbone is expected to house latent capacity that can then be sold to private sector actors permitted to resell the bandwidth to consumers. These rates are to be agreed with the city through service agreements to ensure affordability to end-users (DBSA, 2012). The business model is innovative as it allows for multiple independent operators to sell bandwidth services of almost any scale to end-users, be they corporations or households. Since the network will be linked to various national backbones as well as undersea cables, operators will be able to mix and match the suppliers of data and services to build the kind of network that suits their own business models (DBSA, 2012: 97). What the province aims for is a mesh of networks and services that ensure access to ICT across the city either through private connection or access at a public facility.

City of Cape Town smart city initiatives

In late 2011, the Swedish company Siemens conducted a 'Sustainable Cities Tour' with the intention of showcasing their technologies for sustainable infrastructure management. The company cites many historical examples of its involvement in the City of Cape Town in South Africa: construction of the first telegraph line between the city and Simonstown in 1860 and the first electric cable car in 1927. The tour culminated in a substantial presence at COP 17 in Durban in December 2011 (Siemens, 2011). The particular emphasis of the tour was on showcasing the company's renewable energy technology. Reference was made to the city's Climate Smart Cape Town programme, mainly focused on an online resource and discussion space on climate change. The tour was followed up with a smart city summit in mid-2012, organised by the City of Cape Town and the Swedish embassy. The opening addresses by the Swedish ambassador and the mayor of the city expressed political enthusiasm for the notion of smart cities as vehicles for sustainable development. The Stockholm Royal Seaport was referred to as a model worth aspiring to (City of Cape Town, 2012a). In a blog, local think-tank Future Cape Town (2012) presented a cautiously enthusiastic potential for inter-country collaboration: the potential for technology transfer, the possibility of utilising Swedish innovations for developmental gains and Cape Town thereby becoming a laboratory for smart solutions to climate change elsewhere.

Siemens was quick to point out its shared history with the City of Cape Town and its past technology transfers. Private and digitally enabled involvement in urban management is best illustrated with a case often overlooked in the city's recent history: the reform of its financial management system. In the tumultuous time following local government restructuring in 2000, accompanied by immense political volatility, 321 staff members from 18 city administrations collaborated in building

Cape Town's first single integrated account monitoring and billing system. The enterprise resource planning was done by SAP, the German multinational business management software company. In a presentation by the city's former Chief Financial Officer (CFO) Philip van Ryneveld (2013), the process of 'co-design' with city officials was discussed, revealing some impressive facts. The system handles over 1 million transactions per day; the initial investment was repaid in less than three years (when calculated in public savings as a result of more effective billing systems) and has yielded substantial net savings. The system is showcased internationally as a success story in system integration.

Van Ryneveld (2013) notes the more subtle benefits to city management that illustrate the finer detail of socio-technical relations. Building the project was a symbol of stability at a time when city restructuring and political flux caused immense insecurity and uncertainty. Alignment between administrative structures, business processes and technology enabled integration beyond the financial system. Administrative reorganisation was done simultaneously. The amalgamation of the financial, billing and payment systems provided a predictable focus at a time of political insecurity. This was in contrast with many of the intractable project processes the city has had to engage with. When probed on the flexibility of the system, van Ryneveld admitted to limited scope for system change, feedback loops from consumers and incorporation of less formal local infrastructure management systems. Regardless of its immutability, it is considered best practice in innovative city financial management and, as shown by the city's former CFO, a core actor in the transformation of the city's administration. The SAP system has been quietly enabling effective billing and system integration.

When examining official documentation and questioning city officials on the *actual* smart city policy, a modest and unglamorous picture emerges. It forms part of the city strategic planning programme, with its core focus on social development. According to its manager, Stelzner, Chief Information Officer (CIO) for the City of Cape Town, the city is nevertheless inundated with approaches from the private sector for software and hardware solutions (interview, 2014). The city's management has nevertheless remained committed to a slow unfolding of backbone infrastructure by connecting all public buildings with high-capacity optic fibre. Excess capacity can then be sold to developers and business to enable last-mile connections. This is essentially part of an infrastructure strategy, managed as a network that private actors can buy into with funding gains invested in broader social uplift. The language of the city's ICT strategy[1] (modestly subtitled 'A Strategy for a Smarter City') is one of enablement and, to some extent, 'trickle-down'. More ambitious is the stated objective 'to completely change the way in which society and local government interacts and collaborates to enhance the quality of life and economic opportunities of all citizens" (City of Cape Town IS&T Directorate, n.d.: 3). The plan defines three broad outcomes: improved city administration through technology; improved services to business and citizens; and ICT-enabled social and economic development (City of Cape Town IS&T Directorate, n.d.).

The provision of public access points is noted as one of the ways through which the 'digital divide' can be addressed. The foundation of this approach is the Smart Cape Access Point Project. The project was initiated in July 2002, and entailed making five computers available in six public libraries across the city for public internet use on a pilot basis. By the end of 2002, there were 3,000 registered Smart Cape users. Increasing demand led to expansion to all 97 public libraries in the city, enabled through a $1 million award from the Bill & Melinda Gates Foundation in August 2003. There are currently 170,000 registered users (City of Cape Town, 2014).

The project initially targeted the poorest areas of the city. Users require only a library card to gain access to computers and the Internet, with one staff member per library assigned an administrative role. Published research on the project reports limited stakeholder engagement, high reliance on volunteers and lack of recognition of the added burden this placed on library staff (Chigona et al., 2010). The project is reliant on donor funding, which raises sustainability concerns. One of the many positive impacts of this initiative has been local economic development. Part of the initial smart city strategy was to enable entrepreneurship. In a comprehensive analysis of the project, Valentine (2004) mentions small businesses that started with the assistance of the Smart Cape facilities, teenagers using computers to construct CVs earning pocket money and at least one spin-off enterprise reported in Khayalitcha, a digital cafe focused on business services for local entrepreneurs. Staff members of the City of Cape Town noted the importance of librarians in enabling learning and training, with each library yielding its own stories in relation to local empowerment and income generation (interview, 2014).

The notion of empowerment does feature quite strongly in documentation on the city's ICT strategy, and the operationalisation thereof reflects a strong official commitment to not be swayed by private sector interests. Smart Cape provides an entry point for the 50 per cent of the population with no access to the Internet (StatsSA, 2011). Variables such as local infrastructure, library staff commitment and demand contribute to a differentiated landscape of appropriation. In some cases the project has partnered with other programmes such as the Violence Prevention through Urban Upgrading (VPUU) initiative in Harare, Khayalitcha, in embedding it within a social precinct, augmenting other educational and social infrastructure through clustering. In other parts of Khayalitcha, a concern with more basic infrastructure, basic sanitation, has led to technology-enabled mobilisation.

Smart cities from the bottom up?

> It is time that the City of Cape Town and local authorities throughout the country get used to the idea that the urban poor do not need to wait for experts to provide data on conditions of life in their communities.
>
> *(Cape Times, 2014)*

The engagement with a broader urbanity could be seen as crucial in determining whether smart city initiatives go beyond integrated city government and place marketing. The reciprocal relationship between content provision and consumer is important. Ongoing interactive exchange is mentioned, but it is worth considering whether this can be truly empowering. The predominance of social media signifies a shift to a more decentralised form of e-governance where citizens could contribute content. But it also reveals a new form of oppositional politics, as shown during the Arab Spring. To this end it is worth exploring a case of digital activism in Cape Town, South Africa.

The Social Justice Coalition Cape Town (SJC) is a civil society organisation based in Khayalitcha. It was formed as a response to perceived institutional failure and political inaction during xenophobic attacks in the country's cities in 2008. The issue that occupies a large part of the organisation's agenda is the lack of safe, operational and accessible toilets in Khayalitcha. The lack of sufficient maintenance, the limited numbers of facilities and high number of attacks on women at night in communal sanitation areas, together with the fact that many of the toilets provided do not have doors, have caused great and justified embarrassment to the city administration.

In the relatively short period that the organisation has been in operation, it has grown to 11 active branches and 40 partner organisations (Social Justice Coalition, 2014a). The main focus areas of the SJC are education, policy engagement and community mobilisation. It does this through its onsite presence in Khayalitcha and other locations, but also through social media. There are over 2,000 followers on Twitter, with just fewer than 3,000 followers on Facebook.[2]

The organisation's website includes online petitions, responses to public press releases and links to news articles. YouTube is used to augment campaigns. Twitter and Facebook are used to stay with the facts in real time, as on-the-ground information is shared amongst press monitoring and meeting announcements. The website includes donation information for donors in South Africa and the United States. Collaborative campaigns can be accessed via the site, such as the 'Campaign for Safe Communities', which includes an online petition and multimedia evidence of issues campaigned against.

In a recent op-ed piece, Robins (Cape Times, 2014) provides a nuanced uncovering of the organisation's approach: data-driven activism where information is used to empower locals in order to confront city government on service delivery. A recent skirmish with the City of Cape Town provides a snapshot of the ways in which SJC employs social media to represent and disseminate information. The organisation's social audit of four informal settlements in one of the city's most marginalised areas, Khayalitcha, revealed that the city administration's janitorial system was not working (Social Justice Coalition, 2014b). The report is available online, with an associated YouTube version, and all associated maps and spreadsheets are also downloadable. The City of Cape Town's response to the report was defensive, and included a claim that SJC was using social media platforms to embarrass the city administration (Cape Times, 2014). The SJC responded accordingly, online, supported by the data necessary to augment its position. The transparency of such exchanges and the painstaking attention to operational detail and meticulous description of data collection practices speak to community insight and data literacy, a form of empowerment that reveals what Robins calls a 'governance from below'.

The online dissemination of surveys and reports, as well as links to media from activist organisations, as a counter to municipal evaluations, has proved to be one of the central tasks of the organisation. Using this information to motivate more rigorous upkeep of communal toilets has resulted in revising service level agreements between the City of Cape Town and contractors (Mitchell, 2014). The fact that these audits on sanitation services are updated monthly is important for ongoing operations.

The website and social media are used in day-to-day monitoring. One of the drop-down menus on the website is entitled 'Imali Yethu', a Xhosa phrase for 'our money'. The menu contains pages with places where service delivery contracts are in place for sanitation, refuse removal and policing in Khayalitcha. The detail of each contract is displayed, with the expected frequency and scope of maintenance tasks, the contact person and detail of the company contracted and time frames. The aim is to empower communities with the necessary facts to monitor and share information on service delivery. Thus information is captured offline, through use of cell-phone cameras and/or note taking, and then uploaded at the SJC office. This does lead to a centralisation of functions, given that a small number of SJC staff have access to online media, leading to some operational difficulties and time lapses (Mitchell, 2014).

An important point to probe is what function online activity serves and whether it has a broader mobilisation role. Mitchell (2014) found that online content tended to be one-directional, functioning as an information portal and also serving as an information resource in placing pressure on the municipality to maintain toilet services. The overall aim is to increase monitoring of public spending and accountability. Each page contains the usual Twitter, Facebook and other social media sharing facilities. As a mobilisation strategy, more broadly, SJC's online activity is constrained by limited Internet penetration in Khayalitcha, the relatively high cost of smart phones and entrenched preferences for face-to-face contact and simply talking on the phone. Using radio is preferable to the Internet, which in many ways is seen as unfamiliar, although the local application 'MixIt' (a well-established South African forerunner to 'WhatsApp') is used for two-way communication between SJC staff and core community activists (Mitchell, 2014).

The efficacy of the SJC online strategies appears to be vested in its ability to respond in a timely way to issues using social media and its physical proximity to communities in need. Its mobilisation function is mainly through partnerships with other organisations, such as alternative news source 'GroundUp Media' (Mitchell, 2014). Another function is engagement with city-wide issues in a considered and informed way. This can be illustrated in a forum organised together with the University of Cape Town-based African Centre for Cities, interrogating the potential for design to become an instrument of social justice and urban inclusion in anticipation of the designation of the city as World Design Capital in 2014. The event included inputs from academics, designers and activists at various sites around the city. Its symbolic value as an online and onsite space to interrogate the creative city discourse provided a meaningful counter to the usual notions of innovation.

Essentially, SJC's social media and online campaigns do represent a 'smart city from the bottom up', but mainly as a challenge to city discourses and, more importantly, as a monitoring strategy. These are important functions that have practical impacts, but with limited mobilisation effects as a result of seemingly mundane but important constraints that speak to larger digital divide issues. As an element of the organisation's 'quiet activism' (Cape Times, 2014), it is part of broader governance that may lack the drama of the 'poo protests' but nevertheless represents an alternative to the corporate smart city discourse.

Reflections on smart Cape Town

The intention of this exploratory overview was to reveal contemporary iterations of the smart city idea within the context of Cape Town. Following an overview of the City of Cape Town's ICT strategy, different categories of examples were briefly discussed: public-led smart city strategies, innovations in other sectors and an example of a bottom-up initiative that engages city policy and operations. These examples are used to give tentative responses to questions that emerged from a definitional discussion of the smart city idea. The three trends identified in the review of global smart city trends and associated corporate discourses are used to reflect on these varied initiatives. They are the representation of new forms of urbanity, a focus on normative agendas such as climate change and relationships to other infrastructural forms. Figure 5.1 summarises related qualities of these initiatives.

Both the provincial and local government ICT policies see digital technologies as transformers, perhaps not explicitly as leading to new forms of urbanity, but certainly as enablers of commercial and social transformation. Greater government efficiency and operational efficiency are qualities seen as more important, and also perhaps essential to the urban connectedness (digital or otherwise) that the discourse speaks of. The surveillance trends discernible in Rio de Janeiro are not (yet) observable, but certainly the City of Cape Town's SAP system does function as a city-wide monitoring system, for those living in formal dwellings anyway. SJC's monitoring 'from below' plays a very practical function in ensuring decent sanitation for many communities in Khayalitcha. The focus on the knowledge economy as an important part of the service economy growth of Cape Town in particular, together with more recent emphases on the creative economy, contributes to a repositioning and reimaging of the city. Digital access is absolutely essential to this ongoing repositioning of the city. In fact, the nurturing of small businesses in marginalised areas through Smart Cape is one of its more successful features in enabling the service economy in marginalised areas.

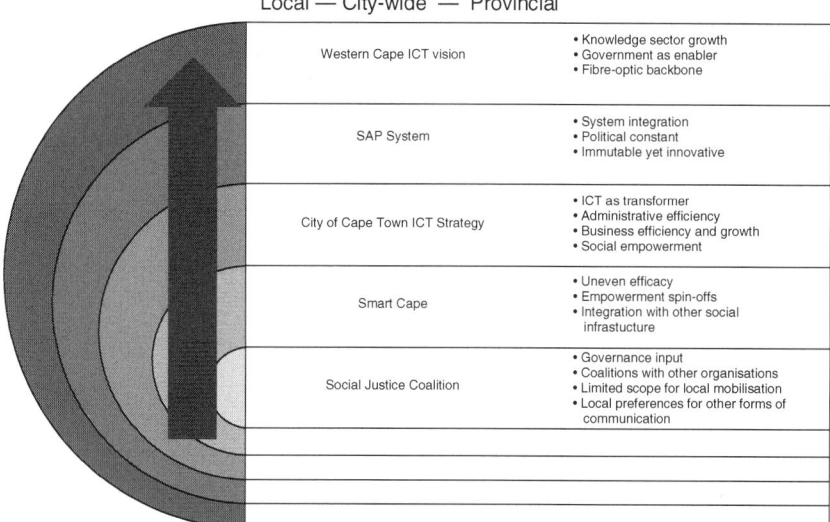

FIGURE 5.1 Smart city initiatives in Cape Town
Source: The author

The emphasis on the climate change agenda and its interface with digital governance, evident in corporate smart city language, is largely absent from the initiatives reviewed here. The Siemens intention to participate in the City of Cape Town's infrastructural transition to more sustainable options is done under the guise of the smart city, but the term's malleability becomes clear when it is used as marketing code for service delivery, sustainability and effective city planning. This initiative is not evident in the city's smart city policy, and appears to be a marketing effort at this point. The language of empowerment and transformation is dominant. When it comes to local empowerment, such objectives are informed by local preferences, contextual factors (such as the commitment of local librarians in the Smart Cape initiative, for example) and infrastructure constraints. The norm of social justice is central in the online and offline activities of the SJC. Here the emphasis is on using information to influence city management, rather than to mobilise. The Smart Cape project reflects a strong commitment to digital access with a strong emphasis on the technology necessary to enable roll-out. Intriguingly, the SAP process in the City during its administrative reform process became and remains a steady, profoundly influential and effective example of digital administration. Its enrolment of important city administration actors, its constancy during difficult times and rapidly proven efficacy were important factors in ensuring a normative constant during a highly volatile political transition.

The relationship to other forms of infrastructure is a feature of these cases but not in the technology-driven language of smart city discourses. Certainly the SAP process is highly centralised and immutable in allowing for billing of utilities. The SJC case reveals the potential malleability of ICT and social media, as it runs a number of campaigns, some with partners, others not, with layers of support that range from onsite assistance to funders in the United States. The end result is ensuring the provision of basic sanitation. ICT is used to pressure local government to fulfil its constitutional mandate. Further research may reveal gaps and issues amongst stakeholders but the efficacy of the organisation clearly lies in its ability to engage issues in real time and enable on-the-ground feedback. The Smart Cape project was criticised for being a central administration idea that was not implemented with enough consultation. Yet research also reveals the creation of local places where connection with other forms of infrastructure provision enables small but important innovations. The library, as knowledge resource and community collective, is an important practical and symbolic access point. In the SJC example, the use of digital technology to mobilise more broadly is constrained by very real infrastructural barriers.

Conclusion

The language of 'smart' in Cape Town is not unlike the narratives that inform the 'real' city. A constitutional mandate to address basic needs and enable social empowerment forms an important backdrop to a range of government-led initiatives that seek to promote technology as an enabler. The role of government as the primary agent in ensuring this is clear. The language of neoliberal economics may be evident in the policy emphases on promoting the knowledge-based economy, but the commitment to social development is as discernible. When 50 per cent of the city's population does not have Internet access, initiatives such as Smart Cape emerge as providing important community entry points. The SJC example shows how the smart city from the bottom up can act as an important governance input in ensuring that local government delivers on its mandate. It may be limited in scope in terms of broader mobilisation and process (at this stage) but, like technology, such situations can change. Right now the SJC and its affiliates use information dissemination as a tool to build relationships with other stakeholders and challenge the City of Cape Town. That in itself speaks of a network urbanism that can thrive within a context where freedom of speech and access to information are fundamental rights. The complexion of the smart city, on the ground, whether enabled through Smart Cape access or the interventions of social movements, is informed by local preferences, barriers and priorities.

Two possibly contradictory trends are discernible and worthy of further exploration. In the Cape Town example, provincial and local government are determined to facilitate access, and partnerships with important private sector actors could potentially yield improved central decision-making and administration. On the other hand, smart phones and social media enable mobilisation and collaboration within civil society. There is a tension, but it is a healthy reflection of the textured nature of contemporary urbanity. Gaining insight into this contributes to an understanding of smart cities from the bottom up, where technology appropriation is more closely linked to livelihoods and mobilisation. Work on this requires taking some of these threads further and understanding technology appropriation more holistically. The small cases explored here show that there are many entry points for doing so.

Notes

1   The ICT strategy is an input into the City of Cape Town's Integrated Development Plan (IDP), a 5-year business plan that coincides with municipal election cycles. It is a continuation, therefore, of the first strategy prepared in 2001, and is due to be revised in 2016 with the next revision of the IDP. This policy context does require that it be aligned with the city's socio-economic goals.
2   Online data last collected in December 2014.

References

Aurigi, A. (2005) Competing urban visions and the shaping of the digital city. *Knowledge, Technology and Society* 18 (1): 12–26.

Bolay, J. and Kern, A. (2011) Technology and cities: what type of development is appropriate for cities of the South? *Journal of Urban Technology* 18 (3): 25–43.

Bovaird, T. and Löffler, E. (2002) Moving from excellence models of local service delivery to benchmarking 'good local governance'. *International Review of Administrative Sciences* 68: 9–24.

Bridges.org (2002) Spanning the Digital Divide. Understanding and Tackling the Issues [online]. Washington, DC. Available at: www.bridges.org/files/active/1/spanning_the_digital_divide [Accessed 14 April 2006].

Business Standard (2014) What on earth is a 'smart city'? [online]. Sunil Seth, New Delhi, 18 July. Available at: www.business-standard.com/article/opinion/sunil-sethi-what-on-earth-is-a-smart-city-114071801449_1.html [Accessed 26 May 2015].

Cape Times (2013a) Poo protests: 'It can't carry on' [online]. Nontando Mposo, Cape Town, 8 August. Available at: www.iol.co.za/news/politics/poo-protests-it-cant-carry-on-1.1559618#.VWSGD1lViko [Accessed 24 January 2015].

Cape Times (2013b) Faeces fly at Cape Town Airport [online]. Zodidi Dano and Clayton Barnes, Cape Town, 26 June. Available at: www.iol.co.za/news/crime-courts/faeces-fly-at-cape-town-airport-1.1537561#.VWSGxVlViko [Accessed 24 January 2015].

Cape Times (2014) Data-driven activism empowering [online]. Steve Robins, Cape Town, 15 December. Available at: www.iol.co.za/capetimes/data-driven-activism-empowering-1.1795657#.VWSB0FlViko [Accessed 24 January 2015].

Caragliu, A., Del Bo, C., Nijkamp, P. (2009) *Smart Cities in Europe*. Research Memoranda Series 0048. Amsterdam: VU University Amsterdam, Faculty of Economics, Business Administration and Econometrics.

Caragliu, A., Del Bo, C. and Nijkamp, P. (2011) Smart cities in Europe. *Journal of Urban Technology* 18 (2): 65–82.

Chigona, W., Roode, D., Nabeel, N. and Pinnock, B. (2010) Investigating the impact of stakeholder management on the implementation of a public access project: the case of Smart Cape. *South African Journal of Business Management* 41 (2): 39–49.

City of Cape Town (2012a) City and Swedish partners to host Smart City development workshop [online]. City of Cape Town media release, 12 June. Available at: www.capetown.gov.za/en/MediaReleases/Pages/CityandSwedishpartnerstohostSmartCityDevelopmentWorkshop.aspx [Accessed 24 April 2014].

City of Cape Town (2012b) City of Cape Town: 2011 Census Data Sheet. Strategic Development Information and GIS Department (SDI&GIS). Cape Town, December [online]. Available at: www.capetown.gov.za/en/stats/Documents/2011%20Census/2011_Census_Cape_Town_Profile.pdf [Accessed 12 January 2015].

City of Cape Town (2014) Smart Cape history [online]. Available at: www.capetown.gov.za/en/smartcape/pages/default.aspx [Accessed December 2014].

City of Cape Town IS&T Directorate (n.d.) City of Cape Town ICT Strategy. Electronic copy received by email.

Cohen, G. and Nijkamp, P. (2002) Information and communication technology policy in European cities: a comparative approach. *Environment and Planning B: Planning and Design* 29(5): 729–755.

DBSA (Development Bank Southern Africa) (2012) The State of South Africa's Economic Infrastructure: Opportunities and Challenges [online]. Available at: www.dbsa.org/EN/prodserv/IIPSA/Pages/default.aspx [Accessed 12 April 2014].

Firmino, R. and Duarte, F. (2010) Manifestations and implications of an augmented urban life. *International Review of Information Ethics* 12 (03): 28–35.

Frost and Sullivan (2014) *Product Leadership Award 2014: Global Best-in-Class Smart City Integrator Visionary Innovation Leadership Award*. Research report. Available at: www.ibm.com/smarterplanet/global/files/us__en_us__cities__FS_IBM_Award_Report.pdf [Accessed 13 July 2014].

Future Cape Town (2012) Smart cities – Swedish solutions to Cape Town's challenges? [online]. Stian Karlsen, Cape Town, 25 June. Available at: http://futurecapetown.com/2012/06/smart-cities-swedish-solutions-to-cape-towns-challenges/#.VWSBPFlViko [Accessed 12 October 2012].

Graham, S. (2010) *Cities Under Siege: The New Military Urbanism*. London: Verso Books.

Hollands, R. G. (2008) Will the real smart city please stand up? Intelligent, progressive or entrepreneurial? *City* 12 (3): 303–320.

IBM (2014) Smarter, More Competitive Cities [online]. Available at: www.ibm.com/smarterplanet/uk/en/smarter_cities [Accessed 10 October 2014].

Misuraca, G. (2007) *E-Governance in Africa: From Theory to Action. A Handbook on ICTs for Local Governance*. Asmara, Eritrea: Africa World Press.

Mitchell, H. (2014) Information and Communication Technologies and Urban Transformation in South African township communities. Unpublished M Phil thesis, University of Cape Town.

Odendaal, N. (2003) Information and communication technology and local governance: understanding the difference between cities in developed and emerging economies. *Computers, Environment and Urban Systems* 27: 585–607.

Odendaal, N. (2006) Towards the digital city in South Africa: issues and constraints. *Journal of Urban Technology* 13 (3): 29–48.

Odendaal, N. (2011) Splintering urbanism or split agendas? Examining the spatial distribution of technology access in relation to ICT policy in Durban, South Africa. *Urban Studies* 48 (11): 2375–2397.

Organisation for Economic Cooperation and Development (OECD) (2008) *Territorial Review: Cape Town, South Africa*. Paris: OECD Publishing. Available at: www.oecdbookshop.org/browse.asp?pid=title-detail&lang=en&ds=&ISB=9789264049635 [Accessed 12 January 2015].

Santinha, G. and De Castro, E. A. (2010) Creating more intelligent cities: the role of ICT in promoting territorial governance. *Journal of Urban Technology* 17(2): 77–98.

Siemens (2011) Inspiring change for a sustainable Cape Town: the Siemens Sustainable Cities tour visits the Mother City [online]. Dale Ladner, Johannesburg, 30 November. Available at:

www.discoversiemensafrica.com/topics/inspiring-change-for-a-sustainable-cape-town/ [Accessed 12 April 2014].

Singer, N. (2012) Mission control, built for cities: IBM takes 'Smarter Cities' concept to Rio de Janeiro. *New York Times*, 3 March. Available at: www.nytimes.com/2012/03/04/business/ibm-takes-smarter-cities-concept-to-rio-de-janeiro.html?_r=0 [Accessed 10 July 2014].

Social Justice Coalition (2014a) About Us [online]. Available at: www.sjc.org.za/about-us [Accessed 2 December 2014].

Social Justice Coalition (2014b) Our Toilets Are Dirty. Report of the Social Audit into the Janitorial Service for Communal Flush Toilets in Khayelitsha, Cape Town: 14–19 July [online]. Available at: nu.org.za/wp-content/uploads/2014/09/Social-Audit-report-final.pdf [Accessed 20 May 2015].

StatsSA (Statistics South Africa) (2011) City of Cape Town Municipal Profile [online]. Available at:www.statssa.gov.za/?page_id=1021&id=city-of-cape-town-municipality [Accessed 24 January 2015].

Valentine, S. (2004) *E-Powering the People: South Africa's Smart Cape Access Project*. Washington, DC: Council on Library and Information Resources.

van Ryneveld, P. (2013) PowerPoint presentation, African Centre for Cities, University of Cape Town, 8 May.

Watson, V. (2014) African urban fantasies: dreams or nightmares? *Environment and Urbanization* 26: 561–567.

WCPG Office of the Premier (2010) Department of Local Government: Five-Year Strategic Plan [online]. Western Cape Government: Office of the Premier. Available at: www.westerncape.gov.za/dept/department-premier/documents/plans/2010 [Accessed 12 April 2014].

Western Cape Economic Development Partnership (2013) OneCape 2040: From Vision to Action [online]. Available at: www.wcedp.co.za/news/onecape-2040-from-vision-to-action [Accessed October 2013].

Western Cape Government (2013) Western Cape Government Broadband Implementation Plan: Integrated Master Plan. Progress Update [online]. Department of Economic Development and Tourism. Available at: http://mybroadband.co.za/news/broadband/91159-western-cape-broadband-initiative-progress.html [Accessed 12 October 2014].

6

PROGRAMMING ENVIRONMENTS

Environmentality and citizen sensing in the smart city

Jennifer Gabrys

Introduction: smart and sustainable cities

Cities that are infused with and transformed by computational processes seem to be the object of continual reinvention. While informational or cybernetically planned cities have been underway since at least the 1960s (Archigram, 1994; Forrester, 1969), proposals for networked or computable cities began to appear as regular features in urban development plans from the 1980s onwards (Batty, 1995; Castells, 1989; Droege, 1997; Gabrys, 2003; Graham and Marvin, 2001; Mitchell, 1995). From designing for the plasticity of urban architecture to envisioning the city as a zone for technologically spurred economic growth, digital city developments have remade urban spaces as networked, distributed and flexible sites for capital accumulation and urban experience.

More recent and commercially led proposals for 'smart cities' have focused on how networked urbanisms and participatory media might achieve 'greener' or more efficient cities that are simultaneously engines for economic growth. Smart-city proponents commonly make the case for the necessity of these developments by signalling toward trends in increasing urbanisation. While cities are centres of economic growth and innovation, they are also, smart-city advocates argue, sites of considerable resource use and greenhouse gas emissions and are therefore seen to be important zones for implementing sustainability initiatives. In these proposals decaying or yet-to-be-built infrastructures are identified as sites of prime smart-city development. Smart cities are presented as a neatly packaged way to meet these generalised challenges, thereby ensuring that future cities – whether retrofitted or new – are more sustainable and efficient than ever before.

Although cities infused by digital technologies and imaginaries are not a new development, the implementation of such measures to achieve sustainability directives under the guise of smart cities is a more recent tactic for promoting digital technologies. In many smart-city proposals, computational technologies are meant to synchronise urban processes and infrastructures to improve resource efficiency, distribution of services and urban participation. Digital technologies, and specifically ubiquitous computing, have become a recurring theme in articulating how sustainable urbanisms might be achieved; yet the intersection of smart and sustainable urbanisms is an area of study that has yet to be examined in detail, particularly in relation to what modalities of urban environmental citizenship are emphasised or even eliminated in the smart city.

This chapter takes up the emergence of the smart city as a sustainable city by looking at one particular case study, the Connected Sustainable Cities (CSC) project developed by MIT and Cisco within the Connected Urban Development (CUD) initiative. The CSC aspect of the project consists of design proposals developed between 2007 and 2008 by William Mitchell and Federico Casalegno in the MIT Mobile Experience Lab working in conjunction with Cisco CUD. The Cisco CUD initiative was a partnership initiated in 2006 in response to the Clinton Global Initiative for addressing climate change. Pairing with eight cities worldwide, from San Francisco to Madrid, Seoul and Hamburg, CUD ran until 2010 and has informed Cisco's ongoing project Smart + Connected Communities, which continues to produce smart-city plans, from development underway in Songdo to proposals to develop a 'Sustainable 21st Century San Francisco' (Cisco, n.d.: online).

Situating this design proposal within a range of smart-city projects that include sustainability in their development plans, I examine how this speculative and early smart-city project proposes to achieve more sustainable and efficient urbanisms through a number of ubiquitous computing scenarios to be adapted to existing and hypothetical cities. The CSC project proposal bears strong resemblances to many smart-city developments still underway and, with its connection to Cisco, one of the primary developers of network architecture for cities, is an influential demonstration of smart-city imaginings. Many of the tools developed through the CUD project consist of planning documents, white papers, eco toolkits, multimedia demonstrations and speculative designs meant to guide smart-city development. As an important but perhaps overlooked part of the process of promoting smart cities, these designs, narratives and documents have played an important role in rearticulating the smart city as a sustainable city. Importantly, however, this chapter focuses on these proposals not simply as *discursive* renderings of cities, but as elements within an urban computational dispositif or apparatus (Foucault, 1980), which performs material–political relations across speculative designs, technological imaginaries, urban development plans, democratic engagements through participatory media and networked infrastructures, many of which are folded into present-day urban development plans and practices, even when the smart city is an ever-elusive project to be realised.

Smart-city plans and designs, as proposed and uncertainly realised, articulate distinct materialities and spatialities as well as formations of power and governance. By considering Foucault's concept of environmentality in this context, I examine the ways in which the CSC project performs distributions of governance within and through proposals for smart environments and technologies. I emphasise this aspect of Foucault's (2008) discussion of environmentality in order to open up and develop further his unfinished questioning of how environmental technologies as spatial modes of governance might alter material–political distributions of power and possible modes of subjectification. Revisiting and reworking Foucault's notion of environmentality not as the production of environmental *subjects* but as a spatial–material distribution and relationality of power through environments, technologies and ways of life, I consider how practices and *operations* of citizenship emerge that are a critical part of the imaginings of smart and sustainable cities. This reading of environmentality in the smart city recasts who or what counts as a 'citizen' and attends to the ways in which citizenship is articulated *environmentally* through the distribution and feedback of monitoring and urban data practices, rather than as an individual subject to be governed.

The primary way in which sustainability is to be achieved within smart cities is through more efficient processes and responsive urban citizens participating in computational sensing and monitoring practices. Urban citizens become sensing nodes – or citizen sensors – within smart-city proposals. This is a way of understanding 'citizen sensing' not as a practice synonymous with 'citizen science' but as a modality of citizenship that emerges through interaction with computational sensing technologies used for environmental monitoring and feedback.

In this context, I take up the proposals for smart cities as developed in the CSC project to ask: what are the implications of computationally organised distributions of environmental governance that are programmed for distinct functionalities and are managed by corporate and state actors that engage with cities as datasets to be manipulated? Which articulations of environmentality emerge within sustainable smart-city proposals and developments when governance is performed through environments that are computationally programmed? And when sensing citizens become operatives within urban computational systems, how might environmental technologies delimit citizen-like practices to a series of actions focused on monitoring and managing data? Might this mean that citizenship is less about a fixed human subject, and more about an operationalisation of citizenship that largely relies on digital technics to become animate?

Environmentality

I take up these questions about transformations in urban process, form and inhabitation in order to analyse in greater detail the ways in which the environmental technologies of ubiquitous computing inform urban governance and citizenship in the smart city. 'Environmentality' is a term I use to describe these urban transformations, which I revisit and rework through a reading of Foucault's unfinished discussion of this concept in one of his last lectures in *The Birth of Biopolitics*. Foucault signals his interest in environmentality and environmental technologies as he moves from a historical to a more contemporary and neoliberal consideration of biopolitics in relation to the milieu or environment as the site of governance. Here, he suggests the subject or population may be less relevant for understanding the exercise of biopolitical techniques, since alterations of environmental conditions may become a new way to implement regulation (2007: 22–3; 2008: 259–61).

Working less with an explanation and more with an open-ended suggestion of what he sees as a growing trend toward *environmental* governance rather than subject-based or population-based distributions of governance, Foucault notes, 'Action is brought to bear on the rules of the game rather than on the players, and finally in which there is an environmental type of intervention instead of the internal subjugation of individuals' (2008: 260). Foucault gestures toward a notion of environmentality where influencing the 'rules of the game' through the modulation and regulation of environments may be a more current description of governmentality, above and beyond direct attempts to influence or govern individual behaviour or the norms of populations. Behaviour may be addressed or governed, but the technique is environmental.

Foucault closes his lecture by indicating that in the following week he would examine in greater detail these questions of environmental regulation. However, he does not develop this strand of thought further, and, instead, his six pages outlining his approach to environmentality are included as a footnote in *The Birth of Biopolitics* lectures (2004, 2008). Consisting more of an unanswered question than a theoretical roadmap, Foucault's discussion of environmentality ranges from a historical analysis of the governing of populations to a consideration of more contemporary modes of governance that may have been unfolding or already underway at the time of his lecture. While his specific concept of environmentality remains a footnote to his discussion of neoliberal modes of governance, it is a provocation for thinking through the effects of the increasing promotion and distribution of computational technologies in order to manage urban environments. In what ways do smart-city proposals for urban development articulate and enact distinctly environmental modes of governance, and what are the spatial, material and citizenly contours of these modes of governance?

The use of the term 'environmentality' that I am developing and transforming based on the biopolitics lectures is rather different from the ways in which it has often been taken up based on Foucault's earlier work, from the making of environmentally aware subjects for the purposes of forest conservation in India (Agrawal, 2005) to the use of environmentality as a term to capture the 'green governmentality' of environmental organisations (Luke, 1999). Environmentality as a concept does offer up ways of thinking about governance toward environmentalist objectives. But it is important to bear in mind the translations that are made across environmentality and *environmentalism*. Foucault's analysis of environmentality does not directly pertain to environmentalism as such, but rather to an understanding of governance through the milieu.[1] In fact, Foucault's interest in environmental modes of governance touches on strategies of 'environmental technology and environmental psychology' (2008: 259), fields that could include designing survival systems or shopping mall experiences (e.g. Anker, 2005; Banham, 1984). Environmental modes of governance are also as likely to emerge from the failure to meet environmentalist objectives. Events such as Hurricane Katrina, as Massumi suggests in his analysis of environmentality, generate distinct modes of crisis-orientated governance that emerge in relation to the uncertainty of climate change – a condition of 'war and weather' that sets in motion a spatial politics of ongoing disruption and response (2009: 154).

Biopolitics 2.0

Foucault's discussion of environmentality, however abbreviated, addresses the role of environmental technologies in governance and in many ways relates to his abiding attention to the milieu as a site of biopolitical management. Biopolitics, or the governing of life, as he analysed it in its late eighteenth- and nineteenth-century formations, was concerned with 'control over relations between the human race, or human beings insofar as they are a species, insofar as they are living beings, and their environment, the milieu in which they live' (2003: 244–5). If we further take biopolitics to include those distributions of power that inform not just *life*, but also *how to live* (2003: 239–45), then how are *ways of life* governed through these particular environmental distributions?

A different formation of biopolitics emerges in the context of environmentality, since biopolitics unfolds in relation to a milieu that is less orientated toward *control over populations* and instead performs through environmental modes of governance and ways of life. In order to capture and examine the ways of life that emerge within the CSC smart-city proposal, I use the term *biopolitics 2.0* (with a hint of irony) to refer to the participatory or '2.0' digital technologies at play within smart cities, and to examine specific ways of life that unfold within the smart city. Biopolitics 2.0 is a device for analysing biopolitics as a historically situated concept, a point that Foucault stressed in his development of the term. The *2.0* of biopolitics captures the situatedness of this term, which includes the proliferation of user-generated content through participatory digital media that is a key part of the imagining of how smart cities are to operate; it also includes the *versioning* of digital technologies through the transition of computation from desktops to environments (Hayles, 2009), whether in the shape of mobile digital devices or sensors embedded in urban infrastructure, objects, and networks – something that is captured by the term 'City 2.0', which circulates as a parallel term to the smart city.

The biopolitical milieu generates material–spatial arrangements in which and through which distinct dispositifs, or apparatuses, operate. The apparatus of computational urbanism can be analysed through networks, techniques and relations of power that extend from infrastructure to governance and planning, everyday practices, urban imaginaries, architectures, resources and more. But this 'heterogeneous ensemble' can be described through the 'nature of the connection' that unfolds across these elements (Foucault, 1980: 194). In his discussions on biopolitics, the apparatus and the milieu Foucault repeatedly suggests that the ways in which *relations* are performed are key to understanding how modes of governance, ways of life and political possibilities emerge or are sustained.

Computational monitoring and responsiveness characterise the 'nature of the connection' across environments and citizens in smart cities. Biopolitical 2.0 relations are performed through the need to promote economic development while addressing impending environmental calamity, conditions characterised by an *urgency* that Foucault critically identifies as being crucial to the historical situation of the apparatus and, consequently, to the operation of biopolitics (1980: 194–5). Within smart-city proposals and projects, cities are presented as urgent environmental, social and economic problems that the digital reorganisation of urban infrastructures is meant to address by increasing productivity while achieving efficiency.

To say that smart cities might be understood through a biopolitics 2.0 analysis is not so much to suggest that digital technologies are simply tools of control as to examine how the spatial and material programs that are imagined and implemented within smart-city proposals generate distinct types of power arrangements and modes of environmentality and entangle urban dwellers within specific performances of citizenship. Smart-city design proposals on one level establish propositions and programs for how computational urbanisms are to operate; but on another level, programs never go according to plan and are never singularly enacted. Environmentality might be advanced by considering smart cities not as the running of code in a command-and-control logic of governing space but as the multiple, iterative and even faltering materialisations of imagined and lived computational urbanisms.

Connected sustainable cities

Working at this juncture of environmental modes of governance, environmental technologies and sustainability as they are operationalised in smart cities, the CSC project within the CUD puts forward a vision for a near future of ubiquitous urban computing orientated toward increased sustainability. The project proposal materials advocate the smart city as the key to addressing issues of climate change and resource shortages, where sustainable urban environments may be achieved through intelligent digital architectures. The CSC design proposals and policy tools, as well as the core visioning document – *Connected Sustainable Cities* (2008), authored by Mitchell and Casalegno – develop scenarios for everyday life enhanced, and even altered, by smart information technologies, which 'will support new, intelligently sustainable urban living patterns' (Mitchell and Casalegno, 2008: 2).

Within the CSC design proposals the technology that most operationalises smart environments and the programmed interactions between city and citizens is ubiquitous computing in the form of 'continuous, fine-grained electronic sensing' through 'sensors and tags' that are 'mounted on buildings and infrastructures, carried in moving vehicles, integrated with wireless mobile devices such as telephones, and attached to products'. Sensor devices are distributed throughout and monitor the urban environment. The continual generation of data provides 'detailed, real-time pictures' of urban practices and infrastructures that can be managed, synched, and apportioned to support 'the optimal allocation of scarce resources' (Mitchell and Casalegno, 2008: 97). Digital sensor technologies perform urban processes as a project of efficiency, where environments are embedded with computational technologies that provide urban management and regulation.

Like many smart-city proposals, the CSC sites are made smart through several common areas of intervention largely orientated toward increasing productivity while enhancing efficiency. A video lays out the rationale for the project and the core areas it addresses, including platforms developed to aid commuting, home recycling, self-managing one's carbon footprint, facilitating flexibility in urban spaces and collaborative decision-making as model areas in which improved efficiency by means of digital connectivity and improved visibility of environmental data may save resources and lower greenhouse gas emissions. While many of the applications envisaged in the proposal are already in use within cities, from electronic bicycle rental schemes to smart meters for managing energy use, the project suggests a further coordinated dissemination of sensor technologies and platforms for achieving more efficient urban processes.

In the CUD project video and CSC design document, urban design and planning proposals take place not necessarily at the scale of the master plan, but rather at the scale of the scenario (Figures 6.1, 6.2, 6.3). From Curitiba to Hamburg, the episodic urban patterns addressed in these designs and policies include urban services, eco-monitoring toolkits and speculative platforms intended to achieve smart and 'seamless' automated living. Yet in many cases the urban interventions take place in a hypothetical city or in a specified city that is rendered sufficiently general as to be receptive to computational interventions within a universalised language of the everyday. In a design scenario sketched out for 'managing homes' in Madrid (Figure 6.1), numerous capabilities are proposed to make homes more efficient. Mobile phones are GPS-enabled to communicate with sensor-equipped kitchen appliances, so that a family dinner may be cooked by balancing location and timing. The home thermostat will similarly sync with GPS and calendars on mobile phones, so that the home is heated in time for the family's arrival. The organisation of activities unfolds through programmed and activated environments so as to realise the most productive and efficient use of time and resources. In the Madrid scenario, monitoring residents' behaviours in detail through sensors and data is essential for achieving efficiency. With this information, environments are meant to become self-adjusting and to perform optimally.

The CSC efficiency initiatives promise to 'streamlin[e] the management of cities', lessen environmental footprints and 'enhanc[e] how people experience urban life' (Mitchell and Casalegno, 2008: 2). By tracking locations and daily activities, smart technologies present the possibility that dinners will self-cook and homes will self-heat. These 'enabling technologies' perform new arrangements of environments and ways of life: 'smart' thermostats couple with calendars, locations and even 'a human body's "bio-signals"' and 'skin temperature and heart rate' may be monitored

FIGURE 6.1 Madrid: managing homes
Source: Mitchell and Casalegno, 2008 (used by permission)

through sensors to ensure optimum indoor temperatures. Similarly, communication with kitchen appliances is proposed to occur through 'Toshiba's "Femininity" line of home network appliances'. These technologies ensure the home will be warm, safe and provided with the latest recipes (Mitchell and Casalegno, 2008: 58–9).

The importance of the everyday as a site of intervention signals the ways in which smart-city proposals are generative of distinct ways of life, where a 'microphysics of power' is performed through everyday scenarios (Deleuze, 1995: 97). Governance and the managing of the urban milieu occur not through delineations of territory, but through enabling the connections and processes of everyday urban inhabitations within computational modalities. The actions of citizens have less to do with individuals exercising rights and responsibilities, and more to do with operationalising the cybernetic functions of the smart city. Participation involves computational responsiveness and is coextensive with actions of monitoring and managing one's relations to environments, rather than advancing democratic engagement through dialogue and debate. The citizen is a data point, both a generator of data and a responsive node in a system of feedback. The program of efficiency assumes that human participants will respond within the acceptable range of actions, so that smart cities will function optimally. Yet programs for efficiency that are multiply distributed will inevitably be multiply enacted across human and more-than-human registers, so that smart bicycles are left in creeks and sensing devices are hacked to surreptitiously monitor domestic environments or intervene in them. This smart-city proposal raises questions as to how these orchestrated ways of life would be actually lived, thereby rerouting programs of efficiency and productivity.

Programming environments

As specifically rendered through smart technologies, the motivating logic of sustainability becomes orientated toward saving time and resources. This in turn informs proposals for how to embed smart technologies within everyday environments in order to ensure more efficient ways of life. Monitoring is a practice enabled by sensors and so it becomes a central activity in articulating the sustainability and efficiency of smart cities. The sensing that takes place in the smart city involves continually monitoring processes in order to manage them. The urban sense data generated through smart-city processes are meant to facilitate the regulation of urban processes within a human–machine continuum of sensing and acting, such that 'the responsiveness of connected sustainable cities can be achieved through well-informed and coordinated human action, automated actuation of machines and systems, or some combination of the two' (Mitchell and Casalegno, 2008: 98). Humans may participate in the sensor city through mobile devices and platforms, but the coordination across 'manual and automated' urban processes unfolds within programmed environments, which organise the inputs and outputs of humans and machines.

'The programmed city' is a speculative and actual project that has been critical to the ongoing development of ubiquitous computing, but which has also demonstrated the complicated and uncertain ways in which programmable environments are realised (Gabrys, 2010: 58 and *passim*). Programming as described in the CSC document has multiple resonances, signalling the architectural sense of programming space for particular activities (cf. Mitchell, 2003) as well as the programming of urban development and policy and the computational programming of environments. Within smart-city proposals, programming of environments is a way in which the 'nature of the connection' within the computational dispositif is performed across a spatial arrangement of digital devices, software, cities, development plans, citizens, practices and more.

The notion of programming, while specific to computation, is further coupled with notions of what the environment is and how it may be made programmable. Some of the early imaginings of sensor environments speculate on how everyday life may be transformed with the migration of computation from the desktop to the environment (Weiser, 1991). While many of these visions are user-focused, environmental sensors also transform notions of how or where sensing takes place to encompass more distributed and nonhuman modalities of sensing (Gabrys, 2007; Hayles, 2009). The programming of environments is perhaps one of the key ways in which 'the milieu' is now best described as 'the environment', since the post-war rise of the term 'the environment' corresponds with more cybernetic approaches to systems and ecology (Gabrys, forthcoming; Haraway, 1991) and with the use of the term 'environment' to describe the computing environment, referring to the conditions in which computation can operate.

A growing body of research in the area of software studies now focuses on the intersection of computation and space, making the point that computing – often in the form of software or code – has a considerable influence on the ways in which spatial processes unfold or even cease to function when software fails (Graham, 2005; Kitchin and Dodge, 2011; Thrift and French, 2002). While software is increasingly informing spatial and material processes, I situate the performativity of software within (rather than above or prior to) the material–political–technical operations of the computational dispositif, since programmability necessarily signals more than the unfurling of scripts that act on the world in a discursive architecture of command-and-control. Software is also not so easily separated from the hardware it would activate (Gabrys, 2011; Kittler, 1995). Instead, as I suggest here, programmability points to the ways in which *computational logics are performed across material–cultural situations*, even at the level of speculative designs or imaginings of political processes (where computational approaches to perceived urban 'problems' may inform how these issues are initially framed *in order to be computable*), while indicating how actual programs may not run according to plan.

The computational articulations of governance and citizenship within the CSC proposals are uncertain indicators for how urban practices might actually unfold, even when processes are meant to be automated for efficiency – but it is exactly the faltering and imperfect aspects of programmed environments that might become sites for political encounters in smart cities. Some smart-city initiatives are finding that the less 'modern' political structures of city councils, for instance, do not make for easily compatible smart-city development contexts. Urban governance may be divided into multiple wards or councils across and through which the seamless flow of data and implementation of digital infrastructures may be complicated or halted. 'Realizing programs of action' within software development 'is complicated and contested', as Mackenzie notes (2005: 88). Code is also not singularly written or deployed but may be a hodgepodge of just-effective-enough script written by multiple actors and running in momentarily viable ways on specific platforms. A change to any element of the code, hardware or interoperability with other devices may shift the program and its effects. When code is meant to reprogramme urban environments, it also becomes entangled in complex urban processes and materialities that inevitably interrupt the simple enactment of scripts.

Programming participation

The infrastructures at play in the CSC vision partially consist of grids and services remade into smart electrical grids, smart transport and smart water. But they also consist of participatory and mobile citizen-sensing platforms through which urban dwellers are to monitor environments and engage with smart systems. Participatory media and environmental devices facilitate this more sustainable city by enabling forms of participation that are compatible with it. The smart infrastructures and citizen-sensing platforms in the CSC project enable monitoring practices, while structuring responses that regulate or recalibrate everyday practices. Sustainable transit options become more viable through the deployment of 'urban citizenship engagement points' (Connected Urban Development, n.d. a: online) that allow for personalised planning of bus routes, carpooling and bicycle rental. Energy contributions may be made at the intersection of smart transit systems or architectural surfaces and mobile monitoring devices. Urban spaces may be easily reconfigured or adapted to allow working and networking in any location at any time, and to facilitate the 'intensification of urban land use'. The way in which these practices are activated occurs across the programs embedded within urban environments and mobile devices. Digitally enhanced infrastructure and citizens are articulated as corresponding nodes, where technologies and strategies for environmental efficiency become coextensive with citizen participation – and 'changed human behaviour' (Connected Urban Development, n.d. a: online).

While additional design scenarios address traffic in Seoul and work-anywhere-anytime proposals for Hamburg, as well as coordinating public transit in San Francisco and using mobile platforms to organise daily health monitoring, one scenario based in an unnamed North American urban location focuses on 'taking personal responsibility' through the narrative of a love contest between two male friends vying for the attentions of an eco-female (Mitchell and Casalegno, 2008: 102). This scenario demonstrates how 'the biggest variable in sustainability' – that is, 'human behaviour' – may be monitored and advanced effectively through ICT applications (Figure 6.2). The male competitors in this scenario engage in logging their daily travel plans online

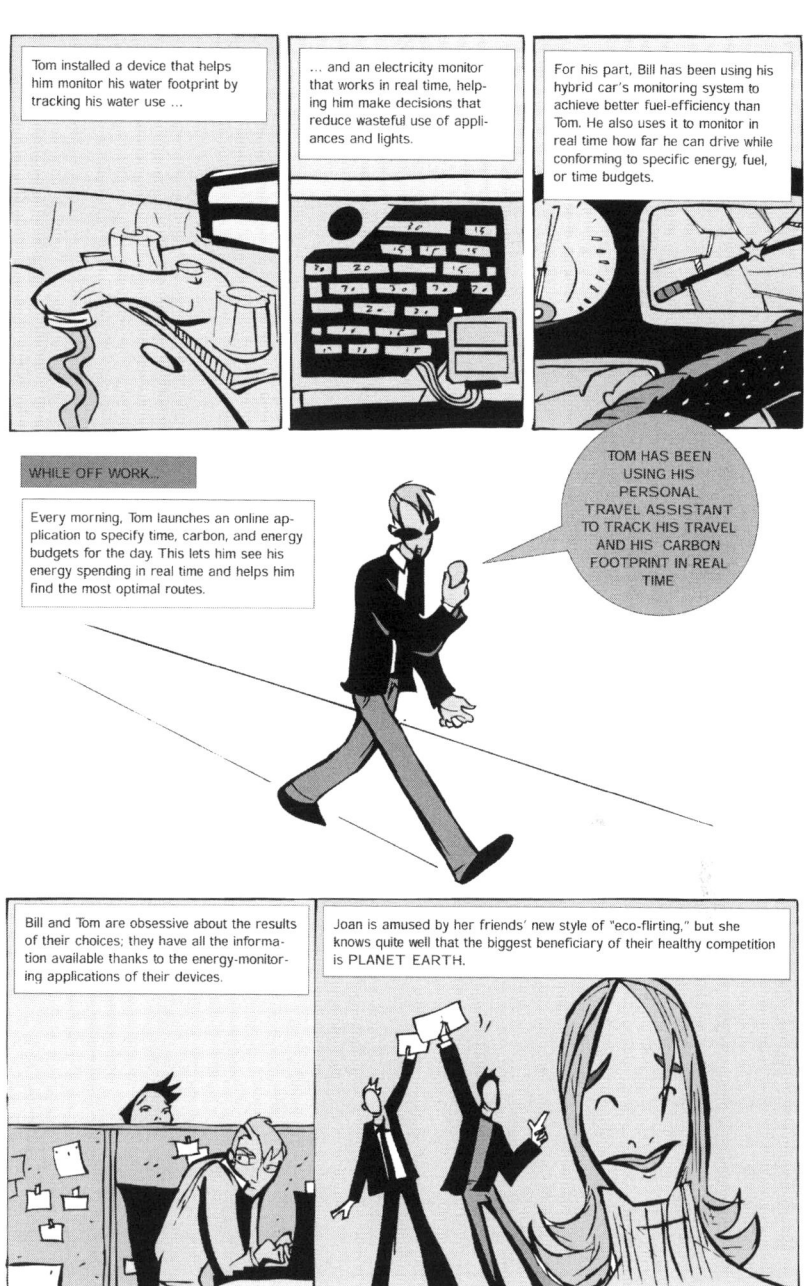

FIGURE 6.2 Enabling technologies
Source: Mitchell and Casalegno, 2008 (used by permission)

to generate carbon footprints for comparison; installing a home monitoring system to measure electricity use; and monitoring water use to generate a water budget. As the scenario outlines:

> Monitor, monitor, monitor … that's a lot of what both men do. They realise that the key to winning Joan's heart is to show her they're making the right decisions, and that means they need a lot of clear information that is meaningful – and actionable.
> (Mitchell and Casalegno, 2008: 89–91)

Monitoring behaviour and generating data is the basis for making sound decisions to advance everyday sustainable practices. Programs of responsiveness are critical to the ways in which sustainable practices are designed to emerge in this smart-city proposal.

In order for these schemes to function, urban citizens need to play their part, whether by taking part in transport systems or by generating energy through their continual movement within urban environments. Urban environmental citizens are responsible for making 'informed, responsible choices' (Mitchell and Casalegno, 2008: 2). Yet these proposals explicitly outline the repertoire of actions and reflections that the smart city will enable, in which the sensing citizen becomes an expression of productive infrastructures. Mitchell and Casalegno stress the benefits of informed participation in urban processes facilitated by participatory media and ubiquitous computing – technologies that, they argue, make a heightened sense of responsibility possible (2008: 101). Urban citizenship is remade through these environmental technologies, which mobilise urban citizens as operatives within the processing of urban environmental data; citizen activities become extensions and expressions of informationalised and efficient material–political practices. Citizens who sense and track their own consumption patterns and local environmental processes have a set of citizen-like actions at their disposal, enabled by environmental technologies that allow them to be participants within the smart city.

The balancing of smart systems with citizen engagement is typically seen as a necessary area to address when considering the issues of surveillance and control that smart cities may generate. As the previously cited Rockefeller-funded report suggests, global technology companies such as IBM and Cisco may have a rather different set of objectives than 'citizen hacktivists', and yet both these companies have vested interests in contributing to emerging smart-city proposals (Townsend et al., 2010). Digital technologies are seemingly liberating tools, allowing citizens to engage in ever more democratic actions; and yet, the monitoring and capture of sensor-data within nearly every aspect of urban life vis-à-vis devices deployed by global technology companies suggest new levels of control. But could it be that this apparent dichotomy between sensing citizen and smart city is less clear-cut? In many ways participatory media could already be seen as tools of variously restricted political engagement (Barney, 2008), while smart urban infrastructures never quite manifest (if they do so at all) in the totalising visions presented.

FIGURE 6.3 Curitiba: citizen reporting
Source: Mitchell and Casalegno, 2008 (used by permission)

The sensing citizen could be seen to be an expression of the ideal mode of citizen participation in smart-city visions, rather than a resisting agent to them. Sensing citizens are the necessary participants in smart cities – where smart cities are the foregone conclusion. Dumb citizens in smart cities would be a totalitarian overshoot, since they would be entities subject to monitoring without participating in the flow of information. The smart city raises additional questions about the politics of urban exclusion, about who is able to be a participating citizen in a city that is powered through access to digital devices. Yet the participatory agency that is embedded within smart-city developments does not settle on an individual human subject, and citizenship is instead articulated through environmental operations. Within the CSC proposals there exists the possibility that given a possible failure or limitation of human responsiveness – a lack of interest in participating in the smart city – the system may operate on its own. In these scenarios, because of a lack of 'human attention and cognitive capacity' as well as a desire not to 'burden people with having to think constantly about controlling the systems that surround them', it may be relevant to deploy 'automated actuation', the project authors suggest. This would mean that urban systems become self- managing such that 'buildings and cities will evolve towards the condition of rooted-in-place robots' (Mitchell and Casalegno, 2008: 98). Citizens might be seen as figures responding within the program of environmentality. However, the smart-city program is able to operate independently by sensing environments as well as actuating them and intervening in them to the point where environmental technologies may override citizens if they do not perform according to pre-set functions – or the rules of the game.

Processes of regulating urban environments within smart-city proposals do not require internal subjugation as such, since governance is distributed within environments that default to automatic modes of regulation. Here is a version of biopolitics 2.0, where monitoring behaviour is less about governing individuals or populations and more about establishing environmental conditions in which responsive (and correct) modes of behaviour can emerge. Environmentality does not require the creation of normative subjects, as Foucault suggests, since the environmental citizen is not governed as a distinct figure; rather, environmentality is an extension of the actions and forces – automaticity and responsiveness – embedded and performed within environments.

Citizen sensing and sensing citizens

A final point of consideration that emerges within smart-city and citizen-sensing frameworks is the extent to which environmental monitoring leads to actionable data. Smart-city infrastructures are projected to operate as a self-regulating environment, but the monitoring technologies that are meant to enable efficiencies within these systems are less obviously able to generate efficiencies or action within 'citizen' practices. In a CSC scenario demonstrating the types of urban environmental citizenship made possible within the green and digital city, proposals are made for residents of Curitiba to experience enhanced and synchronised mass-transit options while monitoring and reporting on air pollution at these nodes (Figure 6.3). Citizen reporting and community engagement are amplified by virtue of ICT connectivity. Through these monitoring and reporting capabilities, positive changes are seen to follow as a result of increased information and connectivity: gather the air pollution data, report to the relevant political body and environmental justice will be realised. These activities and concerns are presented as universally applicable, in that anyone may have cause to monitor and collect pollution data and diligently forward this on to relevant governmental parties. The ambividual actions 'coded' into these processes do not presuppose a particular subject, since a fully automated sensor may equally perform such a function. Rather, these programs of responsiveness allow for a fully interchangeable procession of human-to-machine or machine-to-machine data operations.

A similar trajectory is typically envisaged for self-regulating citizen activities: information on energy consumption will be made visible, a correcting action will be taken and balance to the cybernetic-informational system will be restored. In these scenarios environmental technologies monitor environments and citizens, while citizens monitor environments and themselves. Citizens armed with environmental data are seen to be central democratic operators within these environments. But the 'governing' contained within cybernetics may not neatly translate into the governing of environments (cf. Wiener, 1965). It may be that the very responsiveness that enables citizens to gather data does not extend to enabling them to meaningfully act upon the data gathered, since this would require changing the urban 'system' in which they have become effective operators. Similarly, dominant, if problematic, narratives within sustainability of continued growth through improved efficiency and ongoing monitoring typically do not mobilise an overall resource or waste reduction (what is well known within energy discourse as the 'rebound effect'). Strategies of monitoring and efficiency might be seen to co-opt urbanites into modes of environmentality and biopolitics that leave modes of neoliberal power unexamined, since the aim of realising sustainability objectives through citizen engagement is seen to be a worthy pursuit.

Ultimately, the transformation of citizens in to data-gathering nodes potentially focuses the complexity of civic action toward a relatively reductive if legible set of actions. Participation in this smart and sustainable city is instrumentalised both in terms of remedying environment issues through efficiency, and through devices that will harvest and connect up information to arrive at this outcome. Yet the informational and efficiency-based approach to monitoring environments raises more questions about what constitutes effective environmental action than it answers. In order for such instrumentalisation to occur, urban processes and participation directed toward sustainability in many ways must be programmed to be amenable to a version of (computational) politics that is able to operate on these issues. The modes of sensing as monitoring and responsiveness presented within many sensor-focused and smart-focused cities projects raise the question of whether a 'citizen' might be more than an entity that emerges within parameters of acceptable responsiveness.

Conclusion: from networks to relays, from programs to ways of life

The smart sustainable city vision discussed here is presented as a technical solution to political and environmental issues – an approach that could be seen to be characteristic of many smart-city projects. While the CSC and CUD project proposals are developed as conceptual-level design and planning documents, many of the questions raised here about how smart cities and citizen monitoring projects organise political participation and the imagining of urban environmental citizenship are relevant for considering the proliferation of projects now taking place in these areas, both at the level of community engagement and through urban policy and development partnerships (e.g. European Commission, 2011).

As I have argued, sustainable smart-city proposals give rise to new modes of environmentality as well as biopolitical configurations of governance through distinctly digital dispositifs. Given Foucault's focus on the historical specificity of these concepts and the events to which they refer, it is timely to revisit and revise these concepts in the context of newly emerging smart-city proposals. The environmentality, biopolitics 2.0 and digital political technologies that unfold through many smart-city proposals are expressive of distributions of governance and operations of citizenship within programmed environments and technologies. Such programming is generative of political techniques for governing everyday ways of life, where urban processes, citizen engagements and governance unfold through the spatial and temporal networks of sensors, algorithms, databases and mobile platforms that constitute the environments of smart cities.

The environmentality that emerges through proposals for urban sustainability within the CSC project and many similar smart-city projects involves monitoring, economising, and producing a vision of digitalised economic growth. Such smart cities present ways of life that are orchestrated toward sustainability objectives characterised by productivity and efficiency. The data that develop through these practices are generative of practices of monitoring environments and activities, while activating environmental modes of governance that are located not exclusively within the jurisdiction of 'public' authorities but may also extend to technology companies that own, manage and use urban data. From Google Transit to Cisco TelePresence, HP Halo and Toshiba Femininity, a range of environmental sensor and participatory technologies function in the CSC and other smart-city scenarios that are tools of neoliberal governance, and are operated across state and nonstate actors.

I have emphasised how Foucault's interest in environmentality can be advanced in the context of smart cities to consider how distributions of power within and through environments and environmental technologies are performative of the operations of citizenship – rather than of the individual *subjectness* of citizenship. The 'environmentalist' aspects of the smart and sustainable city are not contingent on the production of an environmentalist or reflexively ecological subjectivity, and the performance of smart urban citizenship occurs not by expanding the possibilities of democratically engaged citizens, but rather by delimiting the practices constitutive of citizenship. The 'rules of the game' of the smart city do not articulate reversals, openings or critiques of urban environmental ways of life. Rather, practices are made efficient, streamlined and orientated toward enhancing existing economic processes.

By pushing Foucault's notion of environmentality even further, I suggest that his concept of the 'rules of the game' might be recast in the context of smart cities less as rules and more as programs – here of responsiveness – that delimit and enable in particular ways, but that also unfold, materialise or fail in unexpected ways. If urban programs are not singular and are continually in process, then environmentality might also be updated to address the ways in which programs do not go according to plan, and work-arounds might also emerge. Such an approach is not so much a simple recuperation of human resistance as a suggestion that programs are not fixed, and that in their unfolding and operating they inevitably give rise to new practices of urban environmental citizenship and ways of life that emerge across human and more-than-human urban entanglements.

This approach to ways of life is important in formulating not a simple denunciation of the smart city, but rather a proposal for how to attend to the distinct environmental inhabitations and modalities of citizenship – and possibilities for urban collectives – that emerge in smart-city proposals and developments. Subjectification, which Deleuze (1995: 83–118) discusses as an important concept in Foucault's work, is ultimately concerned not with the production of fixed subjects, but rather with the possibility of identifying, critiquing and even creating ways of life. Smart-city projects require an attention to – and critique of – the ways of life that are generated and sustained in these proposals and developments. Critique, as articulated in a conversation between Deleuze and Foucault, can be an important way in which to experiment with political engagements and form 'relays' between 'theoretical action and practical action' (Foucault, 1977: 207). From this perspective the ways of life proposed in the CSC scenarios might serve as provocation for thinking through how to experiment with urban imaginaries and practices in order not to be governed *like that*. If we read biopolitics 2.0 as a concept attentive to the ways of life that are generated and sustained within smart cities, and if this computational apparatus operates environmentally, then what new relays for theory and practice might emerge within our increasingly computational urbanisms?

Acknowledgements

The research leading to these results has received funding from the European Research Council under the European Union's Seventh Framework Programme (FP/2007–2013) / ERC Grant Agreement n. 313347, 'Citizen Sensing and Environmental Practice: Assessing Participatory Engagements with Environments through Sensor Technologies'.

Note

1   While the English version of this passage in *The Birth of Biopolitics* translates this term as 'environmentalism', in the French original Foucault uses the term 'environnementalité', which is much closer to conveying the sensing of governmentality distributed through environments, rather than a social movement orientated toward environmental issues (see Foucault, 2004: 266; 2008: 261).

References

Agrawal, A. (2005) *Environmentality: Technologies of Government and the Making of Subjects.* Durham, NC: Duke University Press.

Anker, P. (2005) The closed world of ecological architecture. *Journal of Architecture* 10: 527–552.

Archigram (1994) *A Guide to Archigram, 1961–74.* London: Academy Editions.

Banham, R. (1984) *Architecture of the Well-tempered Environment: Theory and Design in the First Machine Age.* Chicago: University of Chicago Press.

Barney, D. (2008) Politics and emerging media: the revenge of publicity. *Global Media Journal* 1 (1): 89–106.

Batty, M. (1995) The computable city. *International Planning Studies* 2 (2): 155–173.

Castells, M. (1989) *The Informational City: Information Technology, Economic Restructuring, and the Urban–Regional Process.* Oxford: Blackwell.

Cisco (n.d.) Smart + Connected Communities [online]. Available at: www.cisco.com/web/strategy/smart_connected_communities.html [Accessed 9 November 2014].

Connected Urban Development (n.d. a) Introducing Connected Urban Development [online] Available at: www.connectedurbandevelopment.org/ [Accessed 9 November 2014].

Connected Urban Development (n.d. b) Connected urban development visions from MIT's Mobile Experience Lab [online video]. Available at: www.connectedurbandevelopment.org/ [Accessed 9 November 2014].

Deleuze, G. (1995) *Negotiations: 1972–1990*, trans. M. Joughin. New York: Columbia University Press.

Droege, P. (ed.) (1997) *Intelligent Environments: Spatial Aspects of the Information Revolution.* Amsterdam: Elsevier.

European Commission (2011) Report of the Meeting of Advisory Group. ICT Infrastructure for Energy-Efficient Buildings and Neighbourhoods for Carbon-Neutral Cities. Available at: http://ec.europa.eu/information_society/activities/sustainable_growth/docs/smart-cities/smart-cities-adv-group_report.pdf [Accessed 9 November 2014].

Forrester, J. W. (1969) *Urban Dynamics.* Cambridge, MA: MIT Press.

Foucault, M. (1977) Intellectuals and power: a conversation between Michel Foucault and Gilles Deleuze. In *Language, Counter-memory, Practice*, trans. D. Bouchard and S. Simon. Ithaca, NY: Cornell University Press, pp. 205–217.

Foucault, M. (1980) The confession of the flesh. In *Power/Knowledge: Selected Interviews and Other Writings, 1972–1977*, trans. C. Gordon, L. Marshall, J. Mepham and K. Soper. New York: Vintage Books, pp. 194–228.

Foucault, M. (1997) *The Politics of Truth*, trans. L. Hochroth and C. Porter. Los Angeles: Semiotext(e).

Foucault, M. (2003) *Society Must Be Defended*, trans. D. Macey. London: Penguin.

Foucault, M. (2004) *La naissance de la biopolitique. Cours au Collège de France 1978–1979.* Paris: Éditions du Seuil/Gallimard.

Foucault, M. (2007) *Security, Territory, Population: Lectures at the Collège de France 1977–1978*, trans. G. Burchell. New York: Palgrave Macmillan.

Foucault, M. (2008) *The Birth of Biopolitics: Lectures at the Collège de France 1978–1979*, trans. G. Burchell. New York: Palgrave Macmillan.
Gabrys, J. (2003) Cité multimédia: noise and contamination in the information city. Presented at Visual Knowledges conference, University of Edinburgh, 17–20 September. Available at: webdb.ucs.ed.ac.uk/malts/other/VKC/dsp-abstract.cfm?ID=84; http://www.jennifergabrys.net/wp-content/uploads/2003/09/Gabrys_InfoCity_VKnowledges.pdf [Accessed 9 November 2014].
Gabrys, J. (2007) Automatic sensation: environmental sensors in the digital city. *The Senses and Society* 2 (2): 189–200.
Gabrys, J. (2010) Telepathically urban. In A. Boutros and W. Straw (eds) *Circulation and the City: Essays on Urban Culture*. Montréal: McGill-Queens University Press, pp. 48–63.
Gabrys, J. (2011) *Digital Rubbish: A Natural History of Electronics*. Ann Arbor, MI: University of Michigan Press.
Gabrys, J. (forthcoming) *Program Earth: Environmental Sensing Technology and the Making of a Computational Planet*. Minneapolis: University of Minnesota Press.
Graham, S. (2005) Software-sorted geographies. *Progress in Human Geography* 29 (5): 562–80.
Graham, S. and Marvin, S. (2001) *Splintering Urbanism: Networked Infrastructures, Technological Mobilities and the Urban Condition*. London: Routledge.
Haraway, D. (1991) *Simians, Cyborgs and Women: The Reinvention of Nature*. New York: Routledge.
Hayles, N. K. (2009) RFID: human agency and meaning in information-intensive environments. *Theory, Culture & Society* 26 (2–3): 47–72.
Kitchin, R. and Dodge, M. (2011) *Code/Space: Software and Everyday Life*. Cambridge, MA: MIT Press.
Kittler, F. A. (1995) There is no software [online]. *CTHEORY*. Available at: www.ctheory.net/articles.aspx?id=74 [Accessed 9 November 2014].
Luke, T. W. (1999) Environmentality as green governmentality. In É. Darier (ed.) *Discourses of the Environment*. Oxford: Blackwell, pp. 121–151.
Mackenzie, A. (2005) The performativity of code: software and cultures of circulation, *Theory, Culture & Society*, 22: 71–92.
Mackenzie, A. (2010) *Wirelessness: Radical Empiricism in Network Cultures*. Cambridge, MA: MIT Press.
Massumi, B. (2009) National enterprise emergency. *Theory, Culture & Society* 26 (6): 153–185.
Mitchell, W. J. (1995) *City of Bits: Space, Place, and the Infobahn*. Cambridge, MA: MIT Press.
Mitchell, W. J. (2003) *Me++: The Cyborg Self and the Networked City*. Cambridge, MA: MIT Press.
Mitchell, W. J. and Casalegno, F. (2008) *Connected Sustainable Cities*. Cambridge, MA: MIT Mobile Experience Lab Publishing. Available at: http://connectedsustainablecities.com/downloads/connected_sustainable_cities.pdf [Accessed 9 November 2014].
Thrift, N. and French, S. (2002) The automatic production of space. *Transactions of the Institute of British Geographers* 27 (3): 309–335.
Townsend, A., Maguire, R., Liebhold, M. and Crawford, M. (2010) *A Planet of Civic Laboratories: The Future of Cities, Information, and Inclusion*. Palo Alto, CA: Institute for the Future.
Weiser, M. (1991) The computer for the 21st century. *Scientific American* 265: 94–104.
Wiener, N. (1965) *Cybernetics: Or Control and Communication in the Animal and the Machine*. Cambridge, MA: MIT Press.

7
SMART CITY INITIATIVES AND THE FOUCAULDIAN LOGICS OF GOVERNING THROUGH CODE

Francisco R. Klauser and Ola Söderström

Recent urban policy debates have been heavily influenced by discourses reiterating the promises associated with 'smart' information technologies in terms of optimising the management-at-a-distance of urban infrastructures. In Switzerland, as elsewhere, numerous IT-based smart initiatives are being set in motion, relating to a wide range of services and systems, from electricity grids to public transport and traffic management. One of the many terms used for towns and cities embarking upon such initiatives and developments is 'smart cities'.

Although there is today no consensus regarding how exactly to define the IT-mediated smartness of urban infrastructures (Giffinger *et al.*, 2007; Hollands, 2008; Bell, 2012; Kitchin *et al.*, this volume), or which projects, practices and technologies to subsume under the umbrella term 'smart cities', it is possible to identify at least three interrelated centres of gravity around which most approaches navigate. First, discourses on smart cities emphasise the novel possibilities of generating, gathering and processing data which arise from the digitisation of urban systems in the present-day world. Second, smart city developments are presented as the result of novel possibilities to interconnect and to fuse various types and sources of data relating to various aspects of everyday life. Third, the smartness of cities is frequently set in relation to data analytics, thus approached as the correlative of the increasingly automated management of urban systems. The key point here is software, understood as predefined lines of code that process and analyse data with a view to generating automatic responses (Kitchin and Dodge, 2011; Thrift and French, 2002; see also Chapter 2, this volume).

In sum, smart cities are presented as the object of a wide range of technologically mediated practices of management at a distance, based on orchestrated assemblages of computerised systems that act as conduits for multiple crosscutting forms of data collection, transfer and analysis. At their core, efforts towards smart cities thus imply a world of optimised ordering and regulation that relies fundamentally on the coding of social life into software (Haggerty and Ericson, 2000; Lyon, 2007). In other words, smart cities subsume a heterogeneous range of techniques and efforts aimed at governing through code.

Resulting from a two-year research project focused on smart technology applications in the fields of traffic and electricity management, this chapter contributes to contemporary smart city debates in a very specific conceptual and empirical way. Building on Michel Foucault's approach to power and governmentality, and drawing upon empirical insight provided by case studies of two projects relating to smart energy management in Switzerland (iSMART and Flexlast), the chapter explores the internal logics and dynamics of software-mediated techniques of regulation and management at a distance of urban systems. Our key questions are as follows: what power and regulatory dynamics do contemporary smart city developments imply? And how do smart information technologies intervene in the governing of everyday life? Deploying in particular Foucault's concept of 'security' as an analytical heuristic, the chapter approaches these questions on three broad levels; namely, how contemporary governing through code relates to its referent object (referentiality axis), to normalisation (normativity axis) and to space (spatiality axis).

To lay the grounds for this analysis, we first explain briefly the two case studies addressed in the chapter and then move on to outline in some more detail the conceptual approach pursued.

Empirical approach

The two smart-energy projects that will be explored empirically in this chapter are iSMART and Flexlast. The iSMART project constitutes a flagship project for Switzerland. It is devoted not only to the development of novel answers to the technical and organisational issues surrounding the introduction of smart electricity meters, but also to the study of customer behaviours and needs associated with the meters (BKW, 2009: 33). As part of the project, 300 households in Ittigen – a municipality of 11,000 inhabitants, near the city of Berne – were equipped with smart meters and a mobile device (an IP phone with integrated multimedia services). This enables BKW, the electricity provider in the canton of Berne, to study the participants' uses, perceptions, and experiences of this new way of monitoring and managing electricity consumption. Since 2012, two additional projects have been incorporated into iSMART: PowerVISU (aimed at the visualisation and management at a distance of domestic photovoltaic installations) and FLEX (allowing domestic hot water tanks to be controlled and heated automatically by software, depending on fluctuations in both people's electricity needs and the availability of electricity).

The Flexlast case study offers an additional level of technological complexity to this discussion. Flexlast uses three refrigerated warehouses owned by the retailer Migros for the storage of thermal energy, which act as a buffer to help balance fluctuations in the availability of renewable energy on the grid. The key challenge of the project is to calculate and model the exact buffer potential of the warehouses at a given time, depending on anticipated storage volume and logistic activity. The energy in the warehouses can be activated as needed, for better supply and demand matching on the grid. Thus, Flexlast constitutes one of the most ambitious pilots in Switzerland in the field of smart electricity grids (Bundesamt für Energie, 2012; IBM, 2012).

Both iSMART and Flexlast are supported and shaped by IBM and BKW, together with other partners. In the case of Flexlast, the Swiss Federal Office of Energy provides the project funding. Our analysis of the two projects draws upon the extensive study of official documents and reports relating to the two projects, combined with twenty-two semistructured, qualitative interviews conducted in 2012–13 with the partners involved.

Conceptual approach

The chapter adopts a Foucauldian conceptual approach to explore the power and regulatory dynamics inherent in contemporary smart city and smart infrastructure initiatives, as illustrated by iSMART and Flexlast. The main reason for this lies in Foucault's governmentality framework, which allows the study of differing apparatuses of power, understood as historically situated ensembles of techniques for organising and regulating the objects and resources of governing (Foucault, 2008: 186). In differentiating, for example, between juridico-legal, pastoral, disciplinary and security types of power, Foucault (2007) offers a metalevel of analysis that moves beyond a mere description of the specific techniques and discursive regimes through which power acts, to focus instead on the crosscutting rationalities that characterise differing modes of power anchored in specific milieux and historical contexts.

More specifically, we here retain in particular Foucault's conceptualisation of (the apparatus of) 'security' as opposed to 'discipline', as a conceptual tool that allows the emphasis and exploration of the intrinsic flexibility of contemporary governing through code (Bauman and Lyon, 2013), in its relation to reality, normalisation and space (Klauser, 2013; Klauser and Albrechtslund, 2014).

This focus is neither meant to imply that contemporary governing through code entails a strictly homogeneous range of techniques in terms of their regulatory logics, nor to suggest that these techniques should be regarded exclusively as the expression and correlative of Foucauldian security. Rather, our key argument is that Foucault's conceptualisation of security offers a powerful analytical heuristic through which to explore some (but not all) of the power dynamics inherent in contemporary governing through code. The chapter thus also lays stress on a range of principles and issues characterising current smart city developments that Foucault neither explored nor foresaw, but which develop his conceptual and historical framework in very interesting ways. In this sense, our analysis contributes not only to the operationalisation but also to the extension of Foucault's approach to governmentality and power, from a viewpoint centred on the problematics of contemporary governing through code.

With this in view, what matters most for our purposes here is to show how and on what levels Foucault approaches the distinctions and variations between discipline and security. We discuss three levels of distinction below, focusing on how Foucault opposes the two apparatuses with regard to (1) the governed reality or referent object of governing (referentiality axis), (2) normalisation (normativity axis) and (3) space (spatiality axis). This tripartite structure does not provide a definitive or comprehensive guide for organising Foucault's wide-ranging power investigations, but merely offers one possible organising framework, which we hope will prove a useful heuristic in the analysis of iSMART and Flexlast that follows.

Referentiality

The first broad level of analysis on which Foucault distinguishes security from discipline concerns how power in the two apparatuses relates to its referent object (referentiality axis). The main questions are as follows: how is the governed reality approached and conceived? How does power relate to the uncertain, which is inherent in the governing of multiplicities?

Whilst Foucault insists that both discipline and security are concerned with governing reality as a multiplicity of activities, objects and people, he argues that they do so from differing perspectives and according to differing a priori principles. Discipline, on the one hand, designates a specific way of managing multiplicities through techniques of individualisation (2007: 12). Thus disciplinary normalisation consists in breaking down a given multiplicity into specific components, as both the locus and referent object of power put into action (2007: 56–7).

Security, in contrast, works on the relationship between components of a given reality, instead of focusing on the singularised entities separately (Foucault 2007: 47). Reality is approached as a relationally composed whole whose components are deciphered in their intertwined articulation, with a view to their coordinated normalisation. What matters is the optimised adjustment of the assembled components of reality depending on and in relation to each other.

Whilst discipline is essentially centripetal in function and telos – i.e. singularising, concentrating and enclosing – security is centrifugal, constantly expanding and aiming to decipher and interlink ever more extensively and intensively approached components of reality. Thus discipline and security imply not only two fundamentally opposed ways of conceiving and analysing different components of reality and relationships between them, but also two fundamentally opposed a priori principles. Discipline starts from an external, pre-established normative model, whilst security proceeds from the internal, decoded 'normalities' of reality, with a view to optimising their interplay (Foucault, 2007: 63). In sum, the relationship of discipline to reality is singularising, essentialist and, in its derivation from a pre-given normative model, absolute. Security, in contrast, adopts a perspective on reality that is pluralising, relational and relativist (in its derivation from the study of the internal, interdependent normalities of a given reality).

Normativity

The second level of analysis relates to the question of how power in the apparatuses of discipline and security relates to normalisation (normativity axis). The normativity axis implies a focus not only on the aims of governing, but also on the logics and conception of normalisation itself. How do discipline and security conceive of the norm, and of the normal? What does this mean for normalisation?

As mentioned previously, discipline starts from a predefined optimal model that is applied rigidly to the entities individualised for normalisation. The apparatus of security, in contrast, lets things happen within the limits of the acceptable, whilst also regulating and monitoring them with a view to the optimisation of reality in its intertwined components. There are three main consequences of this basic stance: first, it follows that security does not postulate a perfect and final reality to be achieved, but a constant process of optimisation derived from and taking place within a given reality, whose aims and conditions are constantly readapted and redefined, depending not only on the ever-changing parameters of reality itself, but also on the shifting context and conditions of regulation (for example, cost calculations, public opinion and availability of novel control techniques). Thus normalisation in the apparatus of security is inherently processual in its aims and functions.

Second, the normative logic of Foucauldian security is fundamentally flexible in its management of reality. The limit of the acceptable is not merely conditioned by a rigid binary opposition between the permitted and the prohibited, but calculated from and adapted to the differential normalities that characterise the governed reality. The question at stake is how to know, regulate, and act upon this reality within a 'multivalent and transformable framework' (Foucault 2007: 20).

Third, if normalisation in the apparatus of security starts from the decoding of reality in its interacting components, this also means that these components are not valued as either good or bad in themselves, but taken to be natural processes that are granted freedom to evolve according to their internal logics and dynamics, within the acceptable limits of the system (Foucault, 2007: 45). For Foucault, security implies a certain level of freedom – broadly conceived as the 'possibility of movement' (2007: 48–9) – as its basic condition (49). Put differently, for Foucault, security designates the regulatory regime inherent in the (liberalist) art of government that aims at the management of freedom, on the basis of the organisation, fixation, and control of those conditions within which freedom is made possible (2008: 63–4). The important point arising here relates to the contextual logic of normalisation in the apparatus of security. Through techniques of control, calculation, incitation, etc., security aims at the establishment of those conditions and limitations within which the components of reality are to be optimised in their entanglements and aligned internal logics. Thus, on the contextual level, security also relies on prohibitive, coercive – in sum, disciplinary – techniques of power.

Spatiality

Foucault also distinguishes between discipline and security 'by considering the different ways in which they deal with and plan spatial distributions' (2007: 56). This geographical side of Foucault has sparked a number of debates over the years, resulting in a sort of 'geo-governmentality school', as Elden and Crampton put it

(2007: 6; see also Crampton and Elden, 2006; Dillon, 2007; Elden, 2001; Huxley, 2008; Philo, 1992). The third broad level of analysis retained here thus relates to the problem of space (spatiality axis). What forms of spatial organisation do discipline and security produce, and, in turn, how does spatial organisation mediate the exercise of power in the two models?

The disciplinary problem of space, for Foucault, is one of enclosure, fixity and internal structuring, following the need to spatially organise and subdivide artificial multiplicities into singularised entities (2007: 17). In *Discipline and Punish* (1977) Foucault explores this spatial rationality with particular reference to the figure of the panopticon as a paradigmatic spatial model of disciplinary power in action (Hannah, 1997). The spatial logic of security, in contrast, is one not of fixed structuring and enclosure but of managing multiplicities as a whole, in their openness and fluidity. 'Spaces of security' (Foucault, 2007: 11) respond to the need to regulate, optimise, and manage circulations 'in the very broad sense of movement, exchange, and contact, as form of dispersion, and also as form of distribution' (64). The aforementioned conception of freedom as the 'possibility of movement' (48–9) thus also has a spatial meaning.

If discipline and security differ in their spatial problematics and functioning – fixity and enclosure versus circulation and openness – they also contrast in their respective conceptions of spatial organisation, with regard to its mediated and mediating relationship with power (Klauser, 2013). In disciplinary governing, on the one hand, spatial organisation is conceived as something that must be constructed anew, starting from a pre-given raw material. The aim is to arrive at a point of perfection at which spatial organisation fully responds to, and in turn enforces, a pre-given optimal normative model (Foucault, 2007: 19). Again, the figure of the panopticon – in its ideal-typical architectural form aimed at normalisation through spatial organisation – offers a powerful example of this.

Security, on the other hand, approaches spatial organisation as something that relies on and derives from the inherent multidimensionality and 'distributedness' of space, to use Nigel Thrift's expression (2006: 140). Here, space is not conceived as a pre-given raw material to be constructed anew, but as a complex 'composite', made of interlocking, overlapping and distributed (i.e. not necessarily co-located) dimensions, which are deciphered and optimised in their interrelations. This demonstrates, on the level of spatiality, the aforementioned centrifugal reflex of security to approach the entangled components of reality ever more extensively and intensively, with a view to their combined governing.

Governing through code in its relation to reality

Having outlined the Foucauldian distinction between security and discipline, we now start our analysis of the power dynamics implied by contemporary smart city initiatives. Our first level of analysis focuses on how the techniques of governing through code inherent to iSMART and Flexlast relate to the managed reality (referentiality axis).

Governing through interrelation

> BKW backs the use of renewables, based on efficient technology solutions. ... Given the volatility of the renewable energy supply chain, a growing need is to be expected for smart load management solutions that allow for the alignment of energy consumption and provision. Typically, a situation of strong winds and low energy demands results in a system imbalance, which is exactly when we would need to switch on further appliances.
>
> *(BKW corporate developer 1[1])*

This quotation, taken from one of our interviews conducted with BKW, reveals the main purpose of iSMART and Flexlast: both projects aim to align the availability of electricity with its consumption, with a view to maintaining the stability of the grid in a context of increased use of renewable energy. In pursuing this ambition, both projects face the same two-sided problematic, related to (1) the intrinsically volatile and distributed generation of renewable energy and (2) the inherent variability of residential and industrial energy consumption. The key challenge is to bring electricity production and consumption, each with its own internal complexities, into line with each other.

To this end, both projects rely on massive efforts of data generation and data analysis. iSMART, on the one hand, involves the digitisation, monitoring and visualisation of individual electricity consumption, the quantification and monitoring of residential photovoltaic power generation and the study of customer perceptions and uses of smart metering techniques (Kaegi *et al.*, 2011). The three fields of reality thus decoded are combined through data analytics. For example, project participants can monitor in real time how much and what type of energy they consume and how much money they save by adjusting their energy use according to the availability of specific energy sources. Furthermore, iSMART relies on interviews conducted by BKW with the project participants, an approach which permits the study of how customers relate to IT-mediated, personalised electricity management. Thus the pilot not only tests the particular modalities and logics of techno-mediated regulation implied by current smart city developments, but also investigates how these modalities and logics of regulation can be adapted to, negotiated with and coproduced by the actual consumers of the service provided. Here, techno-mediated regulation positively embraces the needs and behaviours of the individuals who voluntarily participate in the control and management apparatus which emerges from it.

Flexlast also implies a form of governing through interrelation, aiming to optimise the balance between energy needs in refrigerated warehouses, the availability of solar and wind energy and the overall stability of the grid. To this end, the project combines warehouse sensor data, along with data supplied by Migros's logistics and scheduling systems, real-time energy data from BKW and Swissgrid and even weather forecasts (Glick, 2012; IBM, 2012). The aim is to keep the warehouses at the correct temperature whilst increasing the use of renewables and taking into account energy needs that are dependent on warehouse logistics (for example, open doors for the delivery of goods, building maintenance and employee schedules). Furthermore, since warehouses functioning as thermal storage facilities can conserve energy and release it into the grid, the project is able to use them as a buffer to help balance fluctuations in the availability of solar and wind energy (Bundesamt für Energie, 2012: 6–7; IBM, 2012). Governing through code in the case of Flexlast thus aims to optimise the interplay between the individual scale and energy needs of the warehouse on the one hand and the collective scale and needs of the electricity grid, on the other; both are approached as flexible variables with their own internal normalities and acceptable limits.

Resonating with Foucault's conceptualisation of security, both projects approach reality as an ensemble of intelligible and analysable components understood as the basic entities and conditions of optimised electricity management. Although the level of the singular is instrumental in this apparatus of power in that it forms the starting point from which explanatory patterns (normalities) are derived through data analytics, it is not the referent object of regulation. The key question is this: how can electricity consumption on the household and industrial level, with its internal complexities, regularities, effects and problems, be taken into account within, and in interaction with, the wider context of grid stability, increased use of renewable energy and customer needs and preferences?

Automated and anticipatory governmentality

The regulatory dynamics that characterise iSMART and Flexlast imply *eo ipso* a mode of regulation that aims at the ever more intensive and extensive study of reality, to decipher its internal regularities. We thus find a combined reflex towards ever-increasing data gathering and ever-wider circuits of data flow. As noted by one of our interviewees from IBM, involved in the planning and development of Flexlast:

> Wherever there is data, there is also software for data analytics. There is a clear trend to process ever more data through software and to interconnect ever more systems, ever more widely. Before, there used to be single systems, whereas today, optimisation is based on system integration
> *(Business Development Executive, Smarter Energy, IBM Switzerland)*

Yet while data processing and management are at the very core of iSMART and Flexlast, both projects, ultimately, strive towards the software-driven automation of electricity management. The following illustrates this:

> Putting great effort into operating my dishwasher at night, buying special light bulbs, etc, I may save 10, 20, perhaps 30 Swiss cents a day. That's obviously quite an effort for a small outcome. ... That's when we naturally come to say, 'all of that has to be managed automatically'.
> *(Interview, BKW corporate developer)*

In referentiality terms, the dynamics of automation inherent in contemporary governing through code is of central importance and requires some further elaboration. Thus, below, we discuss in more detail the power dynamics and implications of the increasingly automated governing of everyday life by smart technologies such as those highlighted in our two case studies. In so doing, we move beyond Foucault's conceptualisation of security, which, given the time at which it was written, does not take into account such developments (Graham, 1998, 2005; Kitchin and Dodge, 2011; Thrift and French, 2002).

Whilst automation is relatively modest in the case of iSMART – it is limited to the heating of residential hot water tanks depending on electricity demand – it is far-reaching in the case of Flexlast. The challenge here is to model and predict the warehouses' power requirements at any given time, taking into account warehouse characteristics, expected logistic activity and other variables, thus allowing reduced energy consumption or activating reverse electricity flows during periods of either high demand or low availability of renewable energy. Drawing upon various grid-relevant and warehouse-relevant data sources, the project elaborates computer algorithms that serve as analytical and predictive tools to calculate and model both the potential for and the necessity of peak levelling.

In different ways and at different levels of complexity, both iSMART and Flexlast thus imply a relationship with reality that is at once calculated and calculating. There are two main implications to highlight here. First, automated governing through code induces a temporal dynamics of regulation in which the relationship between past, present and future manifests itself in a specific way: governing relies on predefined codes, derived from the analysis of the past and applied to the present, to anticipate the future (Klauser and Albrechtslund 2014). As stated by Thrift and French, 'software is deferred. It expresses the co-presence of different times, the time of its production and its subsequent dictation of future moments' (2002: 311). Algorithmic governmentality is also, fundamentally, anticipatory governmentality (Amoore, 2007).

Second, governing through code is inherently performative in its relationship to reality. Computer algorithms constitute not only a tool of analysis but also a grammar of action (Galloway, 2004; Kitchin and Dodge, 2011). As a model and technique of analysis, they simplify reality into a legible order (Budd and Adey, 2009: 1369); as a means of automated response, they perform the future through this order. Governing through code is produced by and in turn produces specific classifications and orderings of reality.

One of the important questions that arise here relates to the adequacy of software to approach and govern the internal complexities and dynamics of reality. As Budd and Adey have argued, 'whilst the relationship between software and the simulations they enable is often less than clear, the practice of using models and simulations is often constrained by the computing tools and languages in which they were written, limiting their accuracy and potential application' (2009: 1370). Future research should provide more detailed empirical evidence with regard to how exactly contemporary smart city initiatives aim to address this issue, and the wider implications this has for everyday social life.

Governing through code in its relation to normalisation

In our discussion of iSMART and Flexlast thus far, we have emphasised the reality-derived and relational mode of normalisation that characterises the two projects. To further develop this discussion of how governing through code relates to normalisation (normativity axis), we will take up and empirically address the three (processual, flexible and contextual) normative logics of Foucauldian security that we outlined above.

In the iSMART project, normative targets for modified energy consumption are set, refined and continuously readapted by each participant individually, depending on specific household conditions, goals and progress made at any given time. In line with these moving, flexible and differential targets, participants can choose and schedule when to purchase what kind of electricity and at what price. The system in turn assesses whether targets are met and visualises success, using a traffic-light system (red for missing targets, orange for meeting targets and green for exceeding targets).

This inherently processual and flexible self-management approach resonates with the now myriad gadgets and applications used by individuals for tracking, quantifying and documenting various aspects of everyday life for purposes of self-surveillance and self-optimisation (Albrechtslund, 2013; Klauser and Albrechtslund, 2014). Offering advanced possibilities for analysis, predictions and recommendations, such tools and services are often framed in terms like 'a good life', 'sustainable lifestyle', 'healthy living' and 'individual responsibility'. Importantly, as in the case of iSMART, individuals are free to decide if and how they want to participate. Yet this freedom to decide is informed and governed on many levels and in all kinds of ways as the following shows:

> Our key question is, 'how can we encourage people to change their behaviour?' ... Energy costs are low, and will probably remain low, in comparison with health costs, etc. But there are other incentives [than financial ones]. What if you are awarded a traffic light colour as feedback? One minute you're red, the next you may be orange or green ....That's motivating.
> *(Interview, Business Development Executive, Smarter Energy, IBM Switzerland)*

The traffic-light system and financial incentives mentioned in the quotation above are just two of the regulatory mechanisms associated with iSMART; other ways in which the project guides the participants' energy consumption include information campaigns, advice generated by software or solicited from customer advisers and apps that simulate alternative energy models or measure the energy consumption of specific appliances.

Together, these mechanisms form a mode of regulation that does not work in a disciplinary way (through rigid prohibitions or prescriptions), but that plays on the customer's desire to optimise his or her electricity consumption. Many of these techniques indeed blur the traditional supplier–customer binary in that they depend on the active involvement of the customer, thus favouring and inciting a constant interaction between supplier and customer.

Through iSMART, BKW's interview-based study of customer preferences goes yet one step further, in that it allows the company to study exactly how customers perceive the system, which in turn helps rework the conditions and framework within which self-governing is allowed to develop. iSMART, in this sense, also aims at the fine-grain adjustment of the fixed parameters within which the interplay of energy availability and consumption can be optimised. Mirroring security's relationship to normalisation, iSMART is not only processual and flexible, but also inherently contextual in function and scope.

Flexlast also implies a processual, flexible and contextual logic of normalisation, although the three aspects are articulated in a different way. First, we find again the idea of permanent optimisation, as expressed in the following quotation, relating to the project's smart grid component:

> Smart grids are subject to continuous improvement, which means technology never stops evolving. We're not saying smart city. We say smarter city, which is a process. Getting smarter implies an evolution. One is never smart and one would never have a smart grid. Rather, one is at different stages of this evolution. What matters is to inject ever more intelligence, managing ever more consumers.
> *(Interview, Business Development Executive, Smarter Energy, IBM Switzerland)*

Thus the ambition of Flexlast is not to achieve and then to conserve a perfect reality. Rather, the stated goal of injecting ever more intelligence implies a continuous regulatory dynamics, based on ever more complex calculations and modelling, considering ever more parameters and bringing together ever wider circuits of information flow.

Second, and as expressed by its name, the key ambition of Flexlast is flexibility. There are two levels to highlight here. On the individual level, the buffer potential of the warehouses allows for more flexible management of the buildings' air-conditioning demands. On the collective grid level, the buffer potential offers flexibility to compensate for the variations caused by the inflexible components of the system. Mirroring Foucauldian security, both levels allow for the matching of supply and demand within a flexible 'multivalent and transformable framework' (Foucault, 2007: 20).

Third, Flexlast also implies a contextual logic of normalisation in that it entails the establishment and recognition of those conditions and limitations of the energy system, imposed by nature, technology or political will, which provide the basic parameters within which the interplay between electricity consumption and production can be optimised. Examples include the pre-given characteristics of the electricity grid, the relative inflexibility of warehouse logistics, specific temperature requirements for particular products and all kinds of political stipulations and industrial regulations. This of course raises the important question of who fixes these (legal, material, technological, political, etc.) conditions – i.e. who has the authority to set the 'disciplined context' that circumscribes the field of intervention 'offered' to governing through code? We address this problematic in more detail elsewhere (Söderström *et al.*, 2014).

In sum, both iSMART and Flexlast thus combine two interdependent regulatory regimes in normativity terms. On the one hand, the two projects imply a normative logic of governing that is fundamentally processual and flexible in its functioning, aiming to optimise the interplay between energy supply and demand, rather than to prohibit or to prescribe in rigid and predefined ways the use or supply of electricity at a given time. On the other hand, on a contextual level, governing through code as illustrated by iSMART and Flexlast also implies a disciplinary logic of governing that aims at fixing those parameters within which flexibility is administered and encouraged.

Governing through code thus works through techniques of calculation that not only aim to decipher and align the internal complexities of interrelating fields of reality, but also help ascertain the limits within which the system is confined. The notion of the 'acceptable', acknowledged and calculated in both projects with regard to, for example, customer preferences, logistical needs and political stipulations, testifies to this problematic. Importantly, this notion also lies at the very heart of Foucault's conceptualisation of security:

> Instead of a binary division between the permitted and the prohibited, one establishes an average considered as optimal on the one hand, and, on the other, a bandwidth of the acceptable that must not be exceeded.
> *(Foucault, 2007: 6)*

It thus appears that both iSMART and Flexlast are shaped at their very core by the search for the right balance between flexibility and fixity – i.e. between security and discipline. This also means that the regulatory logics of the two modes of power are not antagonistic, but embody and nourish each other (Foucault, 2007: 107). As Foucault puts it, 'control is no longer just the necessary counterweight to freedom ... it becomes its mainspring' (2008: 67).

Governing through code in its relation to space

As shall be shown in this third analytical section, relating to how governing through code relates to space (spatiality axis), iSMART and Flexlast both pursue a 'spatial problematic of circulation', which again resonates strongly with Foucault's security (2007: 11). There are at least four elements that substantiate this claim.

First, the spatial problematic of circulation inherent to both projects refers to the aspiration to 'get the grids fit for the future', as one of our interviewees from

BKW puts it. This involves, more specifically, an ambition to (1) optimise the fluidity and efficiency of the electricity grid – i.e. to better target and balance electric power transmission and distribution in order to avoid overloads or redundancy, (2) facilitate the connection between points of power generation and consumption and (3) automate the management of energy flows whilst taking into account the specific needs that characterise both offer and demand, on the basis of increased digitisation, analysis and software-driven modelling and prediction.

Second, both projects aim at a more widely distributed network structure, by integrating additional, decentralised energy feed-in points for purposes of increased grid stability and flexibility. Whilst iSMART incorporates additional photovoltaic installations on rooftops to meet electricity demand at the local level, Flexlast allows reverse energy flows from warehouses to feed into the regional grid. In both projects BKW praises the increased role of 'prosumers' (producing consumers – see McLuhan and Nevitt (1972)) in the elaboration of more flexible energy generation and consumption models (BKW, 2011: 18).

Third, in developing novel solutions for bidirectional energy flows on the electricity grid that favour decentralised energy sources, both projects also convey an ambition to differentiate and to positively or negatively discriminate between varying sources and flows of energy, some of which are facilitated and endorsed, whilst others are considered less attractive and are gradually reduced.

Fourth, the problematics of circulation inherent to iSMART and Flexlast also relate to data transfer and communication, as the correlative of the complex organisational and spatial structure of the grid. More specifically, iSMART involves two-way communication between smart meters and home appliances, and between households and BKW's central communication system, as well as subsequent data procession and transfer to web-based mobile devices that allow customers the remote and mobile monitoring and control of their electricity consumption. Flexlast, in its smart grid dimension, also involves a complex architecture of data transfer and data integration, with a view to the automated management of electricity flows to and from the warehouses.

In this sense, both projects also exemplify the increased possibilities that now exist for interconnecting data sources situated on multiple geographical scales, and for processing and analysing in increasingly automated ways the data thus generated (Giffinger *et al.*, 2007: 10; Hollands, 2008; see also Chapter 2, this volume). What we see emerging here is a form of geographically, socially and institutionally distributed agency with regard not only to who generates energy, but also to who can access the data fused and interconnected within the complex 'surveillant assemblages' (Haggerty and Ericson, 2000) underpinning smart electricity management.

We thus find a spatial dynamics that responds to the need to manage and optimise circulations, rather than fixing and enclosing particular places, people, functions and/or objects. Foucault, in his conceptualisation of the apparatus of security, grasps the spatiality of this kind of surveillance with unequivocal clarity:

> The problem is not only that of fixing and demarcating the territory, but of allowing circulations to take place, of controlling them, shifting the good and the bad, ensuring that things are always in movement, constantly moving around, continually going from one point to another, but in such a way that the inherent dangers of this circulation are cancelled out.
>
> (Foucault, 2007: 65)

Conclusion

Our analysis of iSMART and Flexlast in terms of referentiality, normativity and spatiality highlights a number of crosscutting and interdependent characteristics that define the power dynamics of contemporary governing through code. As we have shown, both iSMART and Flexlast imply a constant process of optimisation, aiming to adjust the balance between electricity consumption and production within the limits of the acceptable. Thereby, the aims and conditions of governing are constantly readapted and redefined, depending not only on the ever-changing parameters of the governed spaces of flows themselves, but also on the shifting context and conditions of regulation (such as cost calculations, public opinion and availability of novel control techniques). The regulatory regime hence emerging relies on a mode of normalisation that is not only derived from reality, relative and plural in scope and scale, but also fundamentally flexible in its aims and functioning. Spatially speaking, iSMART and Flexlast accommodate a range of intersecting efforts which aim to manage energy consumption and production as an ensemble of increasingly interconnected, digitised, and 'technologically empowered' (IBM, 2010) systems of connections, processes and flows.

Importantly, as we have shown, flexibility and interconnectivity are also at the very heart of Foucault's conception of security. This concept, we believe, thus offers a promising heuristic tool for the study of the aims and rationalities of power in action in the present-day world of IT-mediated regulation.

We have looked in this chapter at small-scale initiatives because it is there, we believe, that the everyday logics of smart cities can be best understood. By approaching smart cities as 'governing through code', in referentiality, normativity and spatiality terms, we have both tried to provide an analytical frame for the study of present software-mediated forms of governing and suggested that whatever will come after smart cities ('supersmart' cities?) should be considered within a genealogy of such increasingly ubiquitous regulatory regimes.

Of course, there is much more to be done to sharpen and extend this interpretation. Also, it will be of major importance for future research to experiment with yet other conceptual vocabularies and perspectives, in order to grasp the complex power dynamics that characterise contemporary modes of governing through code. Such reflection is indeed fundamental, we believe, if we are to understand how smart technologies affect everyday life, or if we are to debate the opportunities and risks associated with the much acclaimed smart city.

Acknowledgements

We are very grateful to the Swiss State Secretary of Education and Research for the research funding provided that allowed the development of this research. Furthermore, we would like to thank BKW, and in particular Henrik Müller and Daniel Berner, for granting access to the iSMART and Flexlast pilots and for facilitating the interviews conducted with the project partners involved.

Note

1 All quotations taken from the interviews relating to iSMART and Flexlast have been translated from German by the authors.

References

Albrechtslund, A. (2013) New media and changing perceptions of surveillance. In J. Hartley, J. Burgess and A. Bruns (eds) *The Blackwell Companion to New Media Dynamics*. Chichester: Wiley-Blackwell, pp. 311–321.
Amoore, L. (2007) Vigilant visualities: the watchful politics of the War on Terror. *Security Dialogue* 38 (2): 215–232.
Bauman, Z. and Lyon, D. (2013) *Liquid Surveillance*. Cambridge: Polity Press.
Bell, S. (2012) System city: urban amplification and inefficient engineering. In M. Gandy (ed.) *Urban Constellations*. Berlin: Jovis, pp. 71–74.
BKW (2009) *BKW Group Annual Report 2009* [online]. Available at: www.bkw-fmb.ch [Accessed 15 August 2014].
BKW (2011) *BKW Group Annual Report 2011* [online]. Available at: www.bkw-fmb.ch [Accessed 15 August 2014].
Budd, L. and Adey, P. (2009) The software-simulated airworld: anticipatory code and affective aeromobilities. *Environment and Planning A* 41: 1366–1385.
Bundesamt für Energie (2012) *Flexlast: Erzeugung von Sekundär-Regelleistung durch ein dynamisches Lastmanagement bei Grossverbrauchern* [online]. Available at: www.bfe.admin.ch/php/modules/enet/streamfile.php?file=000000010956.pdf&name=000000290722 [Accessed 15 August 2014].
Crampton, J. W. and Elden, S. (2006) Editorial: space, politics, calculation: an introduction. *Social and Cultural Geography* 7 (5): 681–685.
Crampton, J. W. and Elden, S. (eds) (2007) *Space, Knowledge and Power. Foucault and Geography*. Aldershot: Ashgate.
Deleuze, G. (1992) Postscript on the societies of control. *October* 53: 3–7.
Dillon, M. (2007) Governing through contingency: the security of biopolitical governance. *Political Geography* 26: 41–47.
Elden, S. (2001) *Mapping the Present: Heidegger, Foucault and the Project of a Spatial History*. London: Continuum.
Elden, S. and Crampton, J. W. (2007) Introduction. In J. Crampton and S. Elden (eds) *Space, Knowledge and Power: Foucault and Geography*. Aldershot: Ashgate, pp. 1–16.
Foucault, M. (1977) *Discipline and Punish: The Birth of the Prison*, New York: Pantheon.
Foucault, M. (1982) The subject and power. *Critical Inquiry*, 8: 777–795.
Foucault, M. (2007) *Security, Territory, Population*, New York: Palgrave Macmillan.

Foucault, M. (2008) *The Birth of Biopolitics*. New York: Palgrave Macmillan.
Galloway, A. (2004) Intimations of everyday life: ubiquitous computing and the city. *Cultural Studies* 18 (2–3): 384–408.
Giffinger, R., Fertner, C., Kramar, H., Kalasek, R., Pichler-Milanović, N. and Meijers, E. (2007) *Smart Cities: Ranking of European Medium-Sized Cities*. Vienna: University of Technology, Centre of Regional Science. Available at: www.smart-cities.eu/download/smart_cities_final_report.pdf [Accessed 15 August 2014].
Glick, B. (2012) Swiss consortium turns to business analytics to help build smart electricity grid [online]. *Computer Weekly*, 27 September. Available at: www.computerweekly.com/news/2240164018/Swiss-consortium-turns-to-business-analytics-to-help-build-smart-electricity-grid [Accessed 15 August 2014].
Graham, S. (1998) Spaces of surveillant simulation: new technologies, digital representations, and material geographies. *Environment and Planning D: Society and Space* 16: 483–504.
Graham, S. (2005) Software-sorted geographies. *Progress in Human Geography* 29 (5): 562–580.
Haggerty, K. and Ericson, R. (2000) The surveillant assemblage. *British Journal of Sociology* 51 (4): 605–621.
Hannah, M. G. (1997) Space and the structuring of disciplinary power: an interpretive review. *Geografiska Annaler – Series B* 79 (3): 171–180.
Hollands, R. G. (2008) Will the real smart city please stand up? Intelligent, progressive or entrepreneurial? *City* 12 (3): 303–320.
Huxley, M. (2008) Space and government: governmentality and geography. *Geography Compass* 2 (5): 1635–1658.
IBM (2010) *Smarter cities with IBM software solutions* [online]. Available at: ftp://public.dhe.ibm.com/software/ch/de/multimedia/pdf/transcript-smarter-cities-with-ibm-software-solutions-eng.pdf [Accessed 15 August 2014].
IBM (2012) Swiss energy utility and supermarket chain pilot smart grid using renewable energy [online press release]. Available at: www.zurich.ibm.com/news/12/flexlast.html [Accessed 15 August 2014].
Kaegi, E., Berner, D. and Peter, A. (2011) Flexible thermal load management for ancillary services market: experience of Swiss smart grid pilot project. Proceedings of the 21st International Conference on Electricity Distribution, Frankfurt, 6–9 June. Available at: www.cired.net/publications/cired2011/part1/papers/CIRED2011_0481_final.pdf [Accessed 15 August 2014].
Kitchin, R. and Dodge, M. (2011) *Code/Space: Software and Everyday Life*. Cambridge, MA: MIT Press.
Klauser, F. (2013) Through Foucault to a political geography of mediation in the information age. *Geographica Helvetica* 68: 95–104.
Klauser, F. and Albrechtslund, A. (2014) From self tracking to smart urban infrastructures: towards an interdisciplinary research agenda on big data. *Surveillance and Society* 18 (3): 273–286.
Latour, B. (2005) *Reassembling the Social: An Introduction to Actor-Network-Theory*. Oxford: Oxford University Press.
Lyon, D. (2007) *Surveillance Studies: An Overview*. Cambridge: Polity Press.
McLuhan, M. and Nevitt, B. (1972) *Take Today: The Executive as Dropout*. New York: Harcourt Brace Jovanovich.
Mitchell, T. (2002) *Rule of Experts: Egypt, Techno-Politics, Modernity*. Berkeley, CA: University of California Press.
Philo, C. (1992) Foucault's geography. *Environment and Planning D: Society and Space* 10 (2): 137–161.

Söderström, O., Paasche, T. and Klauser, F. (2014) Smart cities as corporate storytelling. *City* 18 (3): 307–320.

Thrift, N. (2006) Space. *Theory, Culture and Society* 23: 139–155.

Thrift, N. and French, S. (2002) The automatic production of space. *Transactions of the Institute of British Geographers* 27 (3): 309–325.

8

GEOGRAPHIES OF SMART URBAN POWER

Gareth Powells, Harriet Bulkeley and Anthony McLean

Introduction

The language and artefacts of smart grids are increasingly becoming part of the landscape of urban electricity provision. These range from state-mandated roll-outs of smart meters to neighbourhood and city-scale innovation projects designed to test the means through which information technologies and new forms of storage, generation and demand management can be integrated in contemporary cities. Yet, despite their significant political, economic and social consequences, research on smart grids to date has focused on their technical components. There has been limited discussion of the social and geographical dimensions of what we argue are in many ways urban processes. We argue that it is through the coming together of urban and national, private and public, social and technical as well as real and imagined geographies in multi-scalar assemblage processes that smart urban energy projects are developed and come to have consequences for communities. In this chapter we examine these processes by drawing on case studies from the UK and USA through which we highlight the different outcomes being created in each context as well as their common elements.

In understanding the city through its relation to energy and infrastructure we draw on the assemblage thinking of writers such as Blok (2012) and McFarlane (2011). We suggest that smart urbanism, with which this volume is concerned, can be understood as a relational process of assemblage in which the found objects of the urban fabric – power cables, gas pipes, buildings, roads and so on – are re-purposed around new governmental rationales. These are combined with the inserted materialities of meters, data loggers, control room dashboards and screens to create newly alert cities which produce and use energy in markedly different ways from what had been normal in the second half of the twentieth century. This process of assemblage is not a simple one however, nor is it one in which all relations are equal. Instead, we show in the chapter that these are processes which seek to establish the conditions for the cultivation of certain dispositions and new socio-technical means for 'conducting the conduct' of the self, family members and in some cases entire communities (Foucault, 1982; Crampton, 2007a: online). This is a key aspect of smart grids as entities being actively assembled through urban politics; the sense in which they are a socio-technical (both material and discursive) apparatus that seeks to alter, 'the nature of the connection that can exist between … heterogeneous elements' (Foucault 1977, cited in Crampton 2007b: online). Furthermore, they represent an urban response to several threats to the current system of energy provision. Foucault argued that this is a common feature of the dispositif, the 'thoroughly heterogeneous ensemble' of governmental apparatus: '[the dispositif] has as its major function at a given historical moment that of responding to an *urgent need*' (Foucault 1977, cited in Crampton 2007b: online; original emphasis). The perceived urgent need in this context comes from the emerging crisis of the grid through what is widely referred to as the 'energy trilemma'; the need to modernise the grid to provide cities with affordable, secure and low-carbon energy.

We argue that the imposition of smart grids on cities and the cultivation of variously socio-technically enabled smart dispositions in citizens creates new lines of urban politics, which in some instances can be overcome through the configuration of new relations, as seen in the air-source heat pump case study described in the final section of the chapter. In other contexts, however, urban inequalities risk being intensified. By focusing attention and resources on the priorities of those most able to influence the alignment of relations in the relational city – those building the smart city – the co-production of the urban energy assemblages is pushing the concerns of others yet further from the foreground. In developing this argument, we show how smart energy projects can intensify the socialisation of various risks and the privatisation of rewards which have been so powerfully shown to be features of contemporary capitalism (Harvey, 2001: 353).

Using two case studies we suggest that the urban figures as a frontier for the development of and experimentation with smart grids, understood as socio-material assemblages gathering in urban contexts around discourses of sustainability, optimisation and security. We argue that smart grids are being operationalised in urban contexts through a set of neoliberal apparatus and techniques by a diversity of actors. While these socio-material configurations exceed the urban in multiple ways, the development of smart grids is intimately connected to cities as sites of intervention for transformations in the ownership, logics and fabrics of power networks.

In making this argument we develop two claims about the development and deployment of smart grids and suggest that their social and political consequences could be quite profoundly different from existing systems of power supply and use, the histories of which we set out in the first part of the chapter. First we suggest that smart grids are being made possible by a governmental double logic in that while some state institutions are rolling further back from privatised energy provision through other institutions and apparatus the state can be seen to be rolling forward. In doing so state institutions are playing new roles in these arrangements and this, we argue, has consequences for the urban as municipal authorities begin to co-produce urban energy networks in new ways. This requires us to move beyond the analytic binary in which economies are either neoliberal or state-managed to consider new roles and responsibilities for state institutions in smart urban assemblages.

Second, we argue that urban smart grids emerge from a tension between the political economy of networked places and the regulated territorial spatiality of the pre-existing energy economy. We refer here, on the one hand, to urban political economies in which local power geometries have shaped the fabric and function of cities, as a result of which communities are variously affecting and being affected by the emergence of smart urban projects. On the other hand, however, these place-based processes occur alongside regional and national energy markets and must connect with structures and flows that exceed the urban in many ways. We are suggesting that the work done to articulate these otherwise disconnected processes is an important feature of smart urban power projects.

Urban as smart energy frontier

Interpretations of what constitutes a smart power network vary considerably, but in general 'smart grid' usually refers to schemes introducing information and communication technologies (ICTs) and use of data into energy networks in order to overcome perceived problems. Most often these are connected to some combination of a powerful drive (whether political, commercial or environmental) to shift away from fossil fuel-based energy services to renewables while at the same needing to reduce or defer the very large investments in network and generating assets required to maintain security of supply. The sense in which a response is needed, and now, comes both from the urgency of several calls to move away from fossil fuels as well as the dilemma facing network managers. The expected switch to renewables has been anything but a switch; indeed it has thus far been more of an addition, meaning that the power system must both remain able to work as it has in recent decades by energising cities with power transmitted from remote power stations while also facilitating the steady emergence of wind and solar generation in the city. The urban arena represents a particular challenge in this regard, and one to which smart has emerged as a 'solution'.

Beyond the hype about the potential of smart grids and their future significance, 'actually existing' smart grids are usually only visible in specific demonstration projects and interventions. In 2013 there were estimated to be more than 200 smart grid projects in operation around the world (Lewis, 2013), and by 2014 the EU had recorded a catalogue of 459 in the EU alone. The majority of smart grid projects at an advanced stage are being developed in urban areas across Western Europe, North America and Australia. That experimental energy projects are taking place in cities is not new. Electricity grids originally emerged as privately developed and owned entities in cities, with their local monopoly protected by the presence of considerable upfront capital costs and a relative abundance of un-wired spaces rather than regulation. Taken together these factors meant that territorial competition was uncommon. Indeed, by 1918, such was the spread of demand and the multiple forms of supply for electrical power that in London alone there were seventy authorities, fifty different types of systems, ten different frequencies and twenty-four different voltages (Butler, 2001) all serving the growing metropolis. A growing social and political call for universal provision, however, led to various rounds of national institutionalisation, regulation and outright ownership changes which led to the early and essential urban geographies of electricity provision becoming relatively invisible if not largely forgotten. This matters in that it leads us to reconsider what we encounter in smart urban energy projects as a return to the city as a test bed for (re-)electrification rather than an entirely novel process. Most centrally we must once again regard this new phase of power provision as a local and geographically variegated phenomenon, just as was witnessed in London and other cities in the nineteenth and early twentieth centuries. We must also become attuned to the outcomes of this process for those on the 'receiving end' (Massey, 1993) of smart grid experiments, with particular streets and communities winning and losing out in ways that are concealed if we regard smart energy as a national-plan-based approach to infrastructure provision.

Furthermore, historically several factors have contributed to the removal of electricity provision from the urban geographic imagination in the UK and US. These have been both political and technical. In the inter-war period in the UK, for example, the Electric Power Supply Committee recommended the appointment of electricity commissions that would divide the country into district boards which would take over power generation and distribution in their area (Biscoe, 2014). By 1926 the Electricity Act sought to integrate the British electricity supply industry by establishing a 132 KV AC synchronous grid under the newly created Central Electricity Board (Butler, 2001). Doing so required the use and eventual codification of alternating current (AC) generation, which was able to use transformers to manipulate voltage at will so that networks could cover much greater distances, in line with and able to serve the national geography of energy governance. Through this socio-technical accomplishment these networks became the robust and long-lasting materialities of a national geography of power production and consumption over the course of the mid- and late twentieth century. This came with a large-few to small-many topology of production and consumption, which enabled growing and densely populated cities to be energised by distant power stations (Patterson, 1999), leaving their populations largely unaware of the electrical metabolism of their everyday lives. As the post-war period progressed, across Europe, North America, Australia and parts of Asia, Africa and Latin America power delivery become firmly established as a feature of national status and a symbolic projection of modernity. This was accompanied by a supply-side logic of expansion and upgrade which 'encouraged the development of large-scale, centralized infrastructure systems of extensive physical networks drawing on increasingly distant natural resources' (Guy et al., 2001: 5), amplifying the distanciated relation between power and urban communities. In many ways the reintroduction of energy generation in the city and the accompanying reconfiguration of the grid is reversing this process with the effect that urban communities are once again being confronted with the materialities and constraints of power.

We argue that in the contemporary context a new phase of grid management is emerging (Ofgem, 2011) which seeks to intensify some of the characteristics of marketised infrastructures that became widespread in the late twentieth century. In important ways, however, it goes beyond the single, simple and spatially one-dimensional narrative of nationally bound markets for energy. Brown has argued that 'this century will see the localisation of energy production' (Brown, 2012: 206), which we see as a process unfolding alongside Graham and Marvin's argument about the emergence of an 'entirely new infrastructural landscape that radically challenges established assumptions that have underpinned the relations between integrated networks and cities' (Graham and Marvin, 2001: 139). Taken together these two processes of localisation and utility integration through ICT are renewing and deepening the relation ship between cities and energy networks through and as a result of smart urban power projects. Evidence of this can be found in the concentration of European smart grid funding allocated to a small number of urban 'hot spots' where collaborations between distribution network operators (DNOs), transmission system operators (TSOs) and universities have been most effective in forming smart grid projects. The places most likely to host smart grid projects are, according to the EU's 2014 review of all smart grid research in Europe, in 'the vicinity of major organisations involved in research, innovation, or managing the national or regional transmission networks (major cities as London, Paris, Brussels, Barcelona, Roma or university centers as Bilbao, Grenoble, Arnhem, Karlsruhe, Copenhagen)' (Covrig et al., 2014: 37). These criteria, of course, mean that urban sites are particularly likely to be arenas for smart grid research.

These criteria point to the ways in which the urban is essential to smart power. While the logic of universal provision of electricity embedded in national energy projects could effectively bypass geographical difference, the logic of smart grids is both dependent upon and constitutive of locally differentiated forms of power production and use. As urban geographies become key to the development of new forms of power provision and changing patterns of electricity use, previously invisible relations between the urban and the network come to matter in material and political terms.

A future based much more heavily on distributed generation dispersed throughout the urban fabric of cities poses serious technical and social challenges both for cities and the grid. Without viable electrical storage technology,[1] 'smart' demand will need to be supply-synced via context-specific technical and cultural devices. These include in-home displays and market mechanisms to reconfigure the ways in which people demand energy around the new smart grid assemblages. This is in stark contrast to the Fordist–Keynesian model of universal power provision of the post-war period but it is also markedly different from the period of liberalisation and unrestrained consumption that followed. In both these previous contexts the supply processes were demand-led through the use of flexible, predominantly fossil-fuelled, power generators which could be ramped up or down to match the demand of cities without ever imposing constraints on energy use. In the smart grid, flexibility must be induced into energy demand to compensate for the increasing inflexibility of renewable, local and non-storable supply.

Smart grid assemblages promise to 'improve both the physical and economic operation of the electricity system by making it more sustainable and robust, more efficient by reducing losses while at the same time offering economic advantages for all stakeholders' (Verbong et al., 2013: 117). Furthermore, in bringing about this transformation in energy provision and use, smart grids also resonate with attempts elsewhere in the governance of cities to leverage 'networked infrastructures to improve economic and political efficiency and enable social, cultural and urban development' (Hollands, 2008: 307). However, delivering on these promises remains challenging. In the following case studies we draw out some of the political and social consequences for cities and their citizens.

Case studies

Before proceeding to develop our two claims regarding the public–private character of smart grids and the tensions between territorial and topological processes therein, we first introduce the two case studies on which the chapter draws.

UK smart grids: the Customer-Led Network Revolution

The Customer-Led Network Revolution (CLNR) project is a partnership between Durham and Newcastle universities, a major UK gas and electricity supplier (British Gas) and North East England and Yorkshire's low voltage distribution network operator (Northern Powergrid). The project is funded through the UK's Low Carbon Network Fund, a centrally managed fund established by the national energy regulator Ofgem to provide incentives and directly fund power system innovation. The project features experiments with a range of social, technical and socio-technical innovations on the low voltage network and in customers' homes. These include conventional and 'smart' heat pumps, time-of-use tariffs, electric vehicle charge points, a range of grid-side electrical energy storage devices (battery storage), photo-voltaic panels and in-home displays in various combinations, each studied as a discrete 'test-cell'. The project collected consumption data from over 10,000 homes and small organisations, each of which was invited to complete an online survey which resulted in 913 valid responses. From these, 186 participants also took part in a qualitative research visit which included a semi-structured interview and an energy tour of their home or premises. Drawing on CLNR, we consider the economic and political geographies of spatially and temporally overlapping and interacting processes already active in UK energy systems. We examine the interactions between efforts to cultivate a nation of 'smart' consumers in UK electricity

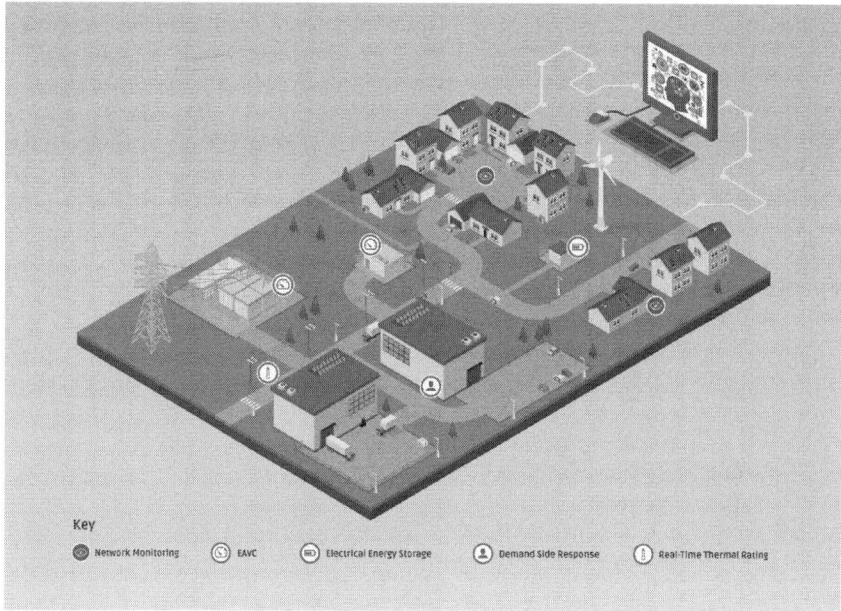

FIGURE 8.1 The low-voltage smart grid as imagined in the UK context
Source: Northern Powergrid © (used by permission)

supply markets and the development of new forms of locally specific urban 'network flexibility'.

Austin and the Pecan Street Project

The Pecan Street Project (PSP) in Austin, Texas, is a non-profit public–private partnership that hopes to provide a 'proving ground' for technologies and ideas that can be used to advocate 'changing the rules, changing the market, providing new incentives [and] educating consumers' (interview, Environmental Defence Fund representative). It is focused on a volunteer group of 1,000 residents and 75 commercial businesses in the city's Mueller district, a planned urban development on a former airport undergoing redevelopment, and operates as a 501(c)3 venture – a non-profit organisation which can attract tax-deductible charitable donations. Although the University of Texas provided an initial $50,000 to kick-start the project, major work did not begin until the US Department of Energy provided a $10.4 million grant in November 2009.[2] This grant money has been matched with $14 million from external partner organisations, mainly private companies, providing funding for research for five years.

In the PSP case study presented here we draw on interviews with a range of actors involved in the project including managers, directors, business owners and

FIGURE 8.2 Solar SunFlowers: a public art installation on the edge of Austin's Mueller district that helps power a nearby retail park
Source: Anthony McLean

administrators from the private, public and voluntary sectors. This is in contrast to the CLNR case study from the UK which draws on interviews with householders.

Drawing on these case studies we develop two arguments about the emergence of smart grids.

Public–private urban smart grid assemblages

The first of our two claims is that the development of smart grids in urban contexts comes with a double logic. In some senses these initiatives can be interpreted as a further intensification of the privatisation of energy provision and marketisation of energy use, through which energy use can be more accurately measured and traded and customers made more flexible and fluid than at any time in the history of networked power provision. While the marketisation of energy systems was accompanied by a rolling back of the state, the creation of these newly sensory networks is being done, however through public–private innovation projects in which local and national state institutions are in some senses rolling forward. State institutions are acting to structure and subsidise projects through mechanisms such as the UK's Low Carbon Network Fund and the US government's 2009 American Recovery and Reinvestment Act to fund energy initiatives, which has led the *New York Times* to call it the 'biggest energy bill in history' (New York Times, 2009: online). At the same time, state institutions are actively directing interactions within smart energy systems. In the UK, for example, a Smart Energy Code has been established as a formal code of conduct that governs all the interactions between parties in the smart power system. Meanwhile a licensed and heavily regulated monopoly contractor has been appointed as the Data Control Company which will perform data flow and access and will curate smart meter data on behalf of customers (who remain the owners of the data) and any party seeking to access their energy use data. Smart grid projects can therefore be seen as co-produced hybrids of state and private interest which emerge from partnerships and which have context-specific social and environmental consequences. This pattern of public–private partnership can be seen in the 2014 review of smart grid projects in the EU in which the 459 projects were found to have a (mean) average of nine partnering organisations per project. Among the parties listed, municipalities, public authorities and the government feature in a list of other, private, entities co-experimenting in mostly urban contexts (Covrig et al., 2014).

We suggest that the emergence of new forms of 'smart' energy provision works by drawing together actors, materialities and communities each with their own political and economic geographies into smart grid coalitions, echoing wider public–private partnerships common in the late 1990s and early part of the twenty-first century in urban and infrastructure governance. Smart grid coalitions, by loosely bundling concerns regarding environment, social justice, urban competitiveness and commercial risk management together in awkwardly inclusive initiatives, have created a consensual, multi-level and multi-stakeholder mode of intervening simultaneously in the city and in the energy network. These developments can be characterised as representing 'socio-technical fixes' (Hodson and Marvin, 2014: 123) and as extensions of the urban entrepreneurialism of the 1990s (Harvey, 1989a; Wood, 1998). Smart grids, we suggest, are one such 'sustainability fix' around which actors and discourses are beginning to establish positions in the urban arena. The effect is that ideas about what it means for the city to be smart are consolidated and made stable through a consensual urban politics of strategic partnerships between elite and or powerful actors such as utilities, universities, housing providers and state institutions.

Public–private heating in Tyneside

To develop these ideas we draw on a trial of domestic air-source heat pumps that was part of the CLNR project. In connection with the CLNR project, 322 air-source heat pumps were installed in homes in the North and North East of England, most of them clustered together on particular distribution network feeders to enable analysis of their combined effects on the distribution network at urban network scale. An air-source heat pump (ASHP) is a device that uses electricity to recover low-level ambient heat energy in the air outside a building for use in the building's space or water heating systems. These devices are typically in the range of 300 to 500 per cent efficient, meaning that for every unit of electricity they use they produce three to five units of heat. Drawing on interviews and home energy tours conducted with eighteen participants from the 322 ASHP users involved in the trial we comment here on the co-produced nature of the trial and what it reveals about the relation between national, regional and local energy economies and places.

Finding hundreds of households in which to install heat pumps was the role of the energy supplier on the project, British Gas (who, despite their name, are the biggest electricity supplier in Great Britain). British Gas operate across the entirety of Great Britain and the design of the CLNR project mirrored the UK's unbundled market structure in that the supplier was responsible for all customer interactions while the DNO was responsible for all network interventions. The ASHP trial, however, required a re-bundling and augmentation of these roles as the trial needed a locally clustered group of customers in order to test the combined local effects of widespread electrification of domestic heat services.

In order to recruit locally concentrated customers to the trial a third-party partner was required, as British Gas's customer base does not follow the urban geography of the network since each household can choose to be supplied by (and thus have a relationship with) any supplier with a licence to operate in the UK. In contrast the DNO has continuous feeder wires that were to be equipped with monitors to sense the combined local effects of the ASHPs at substations. A new entrant to the energy provision system was needed to overcome this mis-match between national-market vs. local-material geographies. This partner was local social landlord South Tyneside Homes (STH), an arms-length management organisation (ALMO) created by South Tyneside Council to manage, maintain and improve its council homes and estates (South Tyneside Homes, 2014). STH is a non-profit-making company that is 100 per cent owned by South Tyneside Council, the municipal government. As an ALMO, STH facilitated access to a housing estate in which all homes were served by the regional DNO and all residents had a relationship with a single housing provider. This became an important relationship that British Gas were able to use to both achieve operational efficiencies but also to manage customer questions, complaints and to provide opportunities for residents to speak to representatives of all CLNR project partner organisations. In doing so, the trial of ASHPs used place-bound relationships created and maintained by municipally owned housing providers to work around and overcome the tensions of the unbundled private energy sector.

A further obstacle was encountered, however, which highlighted the uneven geographies of infrastructure access. Most of the residents recruited to the trial did not have a home internet connection – with the majority either having no internet access at all or using mobile internet only. This meant that the data from the ASHP and domestic circuit monitors would be unable to flow back to the project's data centre. To overcome this issue British Gas had to take on the responsibility of being an Internet service provider to the housing estate, which was both an operational challenge and a novel move for the energy supplier, but was essential in order to make this particular smart grid initiative feasible. In so doing, British Gas crossed over from being a competitive energy supplier to become, in this place and time, a public Internet provider, providing an illuminating example of the blurring of boundaries, roles and responsibilities that are more typical of smart grid assemblages than exceptional features of this case.

We argue that this exemplifies in many ways that the geographies of smart energy do not tessellate neatly and that the tensions between the spatialities of supplier and DNO require place-based partnerships that exceed the vertically imagined unbundled electricity system. In this way these place-based relationships, in which municipal government and its subsidiaries are central actors, become valuable to the grid and lead to local state and quasi-state actors being invited to play a vital role in resolving the tensions within the awkwardly configured privatised power system. We argue that we can see here an example of the co-provision of urban energy governance by both public and private sector actors working in consensual ways to resolve tensions built into the market structure of the UK's energy system. We now turn to a case study from the US to further develop these points.

The public–private smart grid in Texas

> Literally, policemen are laid off if this utility doesn't make a profit.
> *(Interview, Environmental Defense Fund)*

The Pecan Street Project (PSP) reveals that although there are very different drivers in the Texan case, they nonetheless point to the important role played by municipalities in shaping the future of urban energy provision. PSP's status as an independent non-profit allows it to act as an arms-length organisation outside of the control of any single actor, although its founding partners play a key role in directing research. Six organisations have seats on the board – the University of Texas, the City of Austin, the city-owned utility Austin Energy, the Chamber of Commerce, the Environmental Defense Fund and the Austin Technology Incubator (itself a business investment arm of the university). Below this board are a range of external companies who have provided funds and seconded staff to the project.

The high-tech history of Austin and the city's past experience with public–private partnerships has created a group of urban elites who have both accumulated experience of and been acculturated into the doings and sayings (Shove *et al.*, 2012) of public–private collaboration across sectors. In 1982 several US computer and semiconductor manufacturers formed the Microelectronics and Computer Technology Corporation (MCC) in Austin – the first computer industry research and development consortium in the US. This was followed by Sematech in 1986. Both consortia were non-profit research and development organisations and worked in partnership with a diverse range of actors, including state and federal government officials, research institutions, business representatives and manufacturers. For Austin, these consortia were very successful (Smilor *et al.*,

1989) in fostering the conditions for a similar entity to emerge in the form of the Pecan Street smart grid project.

Although the PSP is a public–private partnership, with the city-owned Austin Energy working with a number of private organisations, many participants interviewed believe any smart grid should be private sector-led rather than driven by the state. Distributed generation technologies and demand management systems for sale in an open market were preferred to a mandated state roll-out of smart technologies. Interviewees believed that the change to a new grid should be facilitated by purchases made by willing customers in a competitive market setting. Yet interviewees in Austin also recognised that the deployment of new generation technologies and demand response software could reduce gross demand and therefore the utility's revenue, on which the city relies for the provision of a range of otherwise non-energy-related services, as illustrated by the quotation above. These contingencies between energy, the grid and public services are locally specific, no doubt, but they are part of a wider pattern in smart grid innovation initiatives in which risks and investments are shared by state and commercial entities. Central to this is a common discourse about the future of the grid and a shared rationale that partakes of commercial, environmental and governmental logics which, we argue, is a feature of the wider smart grid assemblages. We suggest that this means that in cities, new constellations of actors will increasingly look for ways to refigure risks and rewards as part of the attempts to connect smart urban futures to the grid. However, while the purpose of smart energy technologies is hybrid in nature – drawing on these multiple rationales – the Austin case reveals that the techniques and apparatuses through which grids are becoming smarter are in many ways neoliberal. They remain based on consumer choice-making and responsibility as the engine for the transformation of energy provision. In so doing they contribute to the normalisation of the view of the urban citizen as an economic agent, as is recognised by Strengers in her depiction of 'resource man' (Strengers, 2013) as the protagonist of smart grid narratives. In this way they contribute to the erosion of other possible narratives of smart urbanism including those of citizenship, sharing and solidarity.

This can be seen in the imagined future of the urban fabric as host to a web of micro- and decentralised power plants in a constant state of transaction. In Austin, researchers are experimenting with a system in which the energy utility is transformed into a socio-technical platform that facilitates peer-to-peer transactions between individual residents generating and consuming locally produced and only locally circulating energy. At thousands of small distributed generation nodes the utility aims to embed metering apparatus to record transactions as well as energy flows in order to artificially construct and record the sale of discrete units of exchange as a means of disentangling an otherwise seamless state of electrical flow and potential. In so doing, the PSP is creating space and socio-technical apparatus for a new energy market to emerge and facilitating individual transactions between urban residents. The utility will operate and maintain the underlying electrical infrastructure – transmission lines, a base generation capacity and an automated software management service – and the new system will charge a subscription fee to those wanting to operate within the decentralised marketplace. In this system, rather than being an energy user, each individual home and business is reconstituted as a 'prosumer' (producer and consumer). One interviewee described this 'brokerage' system:

> I, as a utility operator, am going to be a sophisticated platform that provides energy one way when you need it, takes the energy the other way when you don't need it, monitors the storage and the plug-in and brokers all this distributed onsite generation, storage and consumption. I become the infrastructure, and I take a little fee for transactions for monitoring all this.
> *(Interview, former Austin Energy executive, May 2012)*

The aim is to provide all parties with 'choices and control as opposed to giving the utility or government control' (interview, Environmental Defence Fund representative, May 2012). Thousands of prosumers will engage in constant micro-transactions with peers across the city and what was once a highly centralised, publically managed grid network is imagined to become a dispersed, variegated and dynamic marketplace – yet one still reliant on a large technical network owned and operated by the city. On top of this platform, third parties could develop their own software, hardware and services to sell to residents, while Austin Energy itself will provide a back-up guarantee of service to maintain a basic level of universality to the city. Through this mechanism, the initiative adopts perhaps the most emblematic feature of public–private infrastructure provision and capital expansion: the socialisation of risk and the privatisation of reward.

Actually existing geographies of smart urban power

The second claim we make about smart grids as features of real contemporary cities and those of the near future is that the conjunction of the urban placement of smart grids together with their hybrid political character – which we have discussed above – means that such projects need to be analysed in relation to their particular manifestations in specific places. Informed by Harvey's insight that the urban entrepreneurialism he observed in the late 1980s was driven by a political economy of place rather than territory (Harvey, 1989b), we argue that actually existing smart grids are produced by a tension between the political economy of networked places and the regulated, territorial patchwork of the pre-existing energy economy from which projects are emerging. By networked place we mean both networks in the technical sense – the routes and limits of wires and ICT network coverage – as well as an emphasis on the public–private partnerships – the governance networks – already in place in cities and towns in the vanguard of smart energy experimentation.

In addition to place-based urban interventions, territorially bounded smart networks are also being fashioned through the roll-out of one specific device – the smart meter. A number of smart meter-only projects have already been initiated globally, including in Australia, the UK, where the roll-out involves installing 50 million gas and electricity meters in 27 million homes by 2020, and the United States, where as part of the Recovery Act over 15 million smart meters were installed nationally in each of the last four quarters (Q3 2013 to Q3 2014) (US Department of Energy, 2014). These meters are positioned as necessary preconditions for the creation of smart urban spaces in which consumers are equipped with cultural and technical devices (market structures and metering) which make it possible for energy use and provision to take on new qualities. In more explicitly Foucauldian terms, we are arguing that these devices establish the conditions for new forms of energy use, mentalities and forms of control to be cultivated.

We find it troubling that the socio-technical imaginary of smart energy seems to be a placeless future in which the operating environment is smoothed through the erasure of local, mechanical, physical or contractual boundaries between actors, places and devices. Such devices – what Deleuze and Guattari (Deleuze and Guattari, 2004) might call 'striations' – act as socio-technical markers which divide spaces up, reduce openness and impede flexibility and are notable through their absence from imagined smart futures. Indeed, Hubert observes that 'although they never mention it, the grid must stand as both the emblem and diagram of striated space' (Hubert, n.d.: online). Our contention here is that it is only ever from this highly striated space that smart energy projects start. From these beginnings they seek to establish a hyper-connected set of inter-infrastructures in cities in which nodes, such as homes or businesses, are always connected to everything and follow common protocols which mean also that anything can be plugged into anything. The imagined results are free-flowing data, instantaneous reconfigurability and the avoidance of interruption. The reality of actually existing urban smart grid assemblages is, perhaps inevitably, markedly different, and we find them to be highly striated, uneven and constrained as a result of interactions between the totalising territoriality of smart metering roll-out and the supply markets and the durable geographies of networks and places. In the remainder of this chapter, we use the case studies of Austin, Texas and the North East of England's Customer Led Network Revolution to illustrate these points.

Market territories and grid places

We have argued above that the UK's energy geography comes with in-built tensions as a result of its regionally licensed distribution network monopolies, urban smart grid initiatives and the national-scale-led smart meter roll-out. It is in places, often in cities, where these geographies meet, interact and are negotiated. The UK's national smart meter roll-out, aligned with the geography of the energy supply market, seeks to create a 'smart nation', in which locality and geographic specificities are overcome. It will be the responsibility of energy suppliers, active in the retail market, to buy or rent smart meters and distribute them to customers – passing on the equipment, admin and installation costs to a greater or lesser extent. In basic terms, the immediate benefits of a smart meter are that they provide accurate consumption feedback to the household and accurate meter readings to the energy company, usually every thirty minutes.

Electrical distribution networks are structured differently. These are hierarchies of transformers housed in substations which bring voltage down from ultra-high-voltage national transmission to the 400v low-voltage distribution networks in most streets in the UK. As a result, low-voltage network management and community engagement are conducted by distribution network operators at 'network scale', which is most naturally aligned to 'sub-station communities' – clusters of households and businesses determined by the layout of the wires which result from histories of regulation, network governance and engineering decisions. Motivations for smart grid development at the distribution level are, as a result, focused on enabling the anticipated connection of solar photo-voltaic panels, electric vehicles, electrically powered heat pumps and various forms of energy storage as well as finding ways to avoid or defer investment in network reinforcement. The goal of network trials such as the CLNR is to create local network flexibility as an alternative to costly reinforcement, which had previously been the dominant mode of securing network reliability. Within this logic, contextually specific commercial risk and cost management can be seen to be working in interaction with governmental, regulatorily enforced commitments to renewables and supply security which exert pressure across the regulated space. What results is a rationale for innovation and experimentation through projects which seek to renegotiate these tensions in each instance.

The local is important here for two reasons. First, the local nature of DNO-led smart grids is in contrast to the national geography of supplier-led smart metering roll-out. Second, even locally there is a preference for 'smooth' space, with contiguous customer connections, the absence of boundaries and other 'striations'. The reality is that every house in a street, or every flat in a tower block, will have its own supplier relationship, broadband contract, mobile phone contract and may or may not have other contracts with gas, oil or biomass suppliers. In various ways these variegated states of connectivity amplify, attenuate or prevent customers from adopting an active or smart identity in newly configured urban smart grid assemblages. These striations, the lines between customers of different companies and between formal responsibilities and licences to operate, also dilute the value of customer engagement and in many ways curb the possible contributions of the community to new forms of power provision. Until communities on common feeders begin to act together, by choosing to group-purchase from a common supplier to achieve different forms of energy provision which work with and for them, or until actors coordinate at the local level on behalf of the other stakeholders, citizens will only be able to engage with smart urban power in very diluted ways. This type of community-scale self-awareness and action is of course a rarity.

The new smart marketplace

We now turn to the US case study to consider the implications of new market configurations in urban smart grids. By choosing to use a marketplace as a decision-making and resource-allocation engine the Pecan Street Project system introduces new forms of inequality. We interpret this as a form of power geometry in the way that processes of smart urban power touch down differently in different places and impact communities in different ways – with some benefiting from the opportunities associated with being a prosumer and others being 'on the receiving end' (Massey 1993). For some socio-economic groups Austin Energy will be just one provider of energy management services, with the development of highly individualised and specialised products and contracts to choose from. This is not necessarily a negative aspect of the future smart grid and will be welcomed by many. However, residents with the time and resources will have opportunities and incentives to upgrade their own appliances to improve efficiency, install their own solar panels and wind turbines and then pay Austin Energy to manage their consumption and generation on their behalf. In effect, those able to do so will become players in the market, able to choose which flows to send or receive, which transactions to approve and on which terms to participate. In contrast, those unable to afford the capital investment required to become owners of the still expensive distributed generation technologies could be forced onto flat-rate pay-as-you-go contracts with new and more constraining conditions about home appliance use, albeit through service offerings badged as 'smart' and 'flexible'. In such situations, those configured by rather than configuring the smart grid will be positioned within flows and transactions orchestrated to enhance the positions held by more powerful actors in the marketplace. For example they will be reliant on making their rooftops 'available' to those in the driving seat of smart urban power:

> They'd agree to reduced-cost appliance upgrades such as solar water heaters. They'd participate in Austin Energy's demand response program, which might cycle off their air conditioners in fifteen-minute increments on the city's hottest days. They'd agree to limit their peak use of non-essential appliances in favor of off-peak use. They would never be denied power when they need it. But they would agree that using energy at certain times – outside their service plan – would be 'pay as you go,' just like tossing more garbage than will fit in your city-issued trash can is 'pay as you throw'.
> (Pecan Street Inc., 2010: 16)

As Graham outlined more than a decade ago, such infrastructural 'choice' tends to be limited to 'certain social and spatial groups within the city. The ability to access competing providers is dependent on wealth, location, skills and how lucrative one is to serve' (Graham, 2000: 192). Creating a 'pay as you go' system for those unable to participate in Austin's smart grid will mean that the conditions of possibility for some participants' energy use will be markedly narrower than is currently the case.

The potential for an increase in infrastructural splintering and the targeting of specific socio-economic groups was highlighted by one interviewee:

> [W]e might actually be on the threshold of a word we used to use in the early days, of 'customerisation'. We might actually get to the place where this technology enables the utility to say 'these are stay-at-home moms who keep their air conditioner running and run the dishwasher and have the TV running and a couple of other appliances, and we really ought to figure out a way to keep all of them from being on-peak at the same time'. Go to their house, put these controls in place, stop them from quadrupling their peak for a few minutes at a time. But in my house where my wife and I are both gone all day, don't deploy the hardware ... I would say that it's probably going to be better for us to segment our customers before we try to deploy this crap to every single person.
> *(Interview, Austin Energy executive, May 2012)*

This represents an extension of neoliberal apparatus and techniques into the everyday life of citizens as part of making the public–private smart grid work. While Austin Energy, a state institution, will be rolled back from service provision for urban residents able to be active in the market, it will be simultaneously increasing the scope of their interactions with residents unable to fully become prosumers. In effect this is likely to result in hard-controlling their appliance use and introducing dynamic and time-of-use pricing as 'soft' controls on overall energy demand.

Conclusion

Through these two case studies we have developed an analysis which points to cities and urban contexts as frontier spaces for smart grids and which draws attention to the emergent geographical politics of actually existing, public–private assemblages beginning to deploy smart urban power. Both case studies offer warnings about the difficulties and various flexibilities (on the parts of consumers as well as those bringing the projects forward) needed to reconcile the tensions inherent in the knotted private apparatus of entrepreneurial urbanism and that of regulated power network management. They also illustrate the uneven power geometries at work which produce smart grids – to be experienced differently by different constituencies despite being driven by consensual multi-stakeholder partnerships – unless significant efforts are made to attend to the unevenness that barriers, blockages and inequalities are producing, with clear differences between UK and US experiences. The cases reveal that despite the promise of a smooth space for the development of new products, services, identities and innovations, the reality is that the logic of smart, at least for the short and medium term, is encountering a highly striated urban environment which limits the purchase that logics of smart grid have in terms of their ability to conjure radical energy transitions. This should be a clue to the immanent qualities of smart urban assemblages, that they are being produced by already existing embedded historical energy and urban geographies rather than being inserted into places from elsewhere, or from 'the top'.

The history of urban power outlined earlier stands as a useful reference point and exposes the ways in which urban power provision, smart or otherwise, is always variegated and tied to local economic geographies rather than being rolled out in a uniform manner across territories. However, smart urban assemblages can in many ways be seen as the adoption – by broad coalitions in urban contexts – of the apparatus of the twenty-first century economy, and particularly of its emphasis on flexibility and alertness which resonates so closely with the discourses and practices of smart cities being discussed elsewhere in this book. This is in clear contrast to the dominant mode of infrastructure provision of the twentieth century, outlined in the earlier sections of the chapter, in which network management valued and projected material strength and stability rather than socio-technical flexibility and alertness.

Elsewhere in the chapter we have shown that the use of markets, digital devices and dispositions to create prosumers able to act and react to the 'state' of the grid is a distinctive change in how and from where reliable urban power is provided. It seeds the growth of many tiny nodes throughout the city which each contribute to secure (and hopefully clean) energy provision rather than having cities reliant on a small number of large, remote nodes of generation and control. This comes hand in hand with a change in how citizens demand power. Following from this, we suggest that research on smart urbanism needs to use or develop analytic resources that can attend to the economic geographies of flexibility as they extend out into the everyday lives of citizens as well as how they force new forms of flexibility in terms of how public–private initiatives allocate roles, risks and responsibilities. More particularly, while flexibility has received much attention from geographers in studies of production and labour (Hudson 1989), we argue that geographies of flexible consumption and leisure have been relatively underutilised and have not yet been fully linked to the alertness and responsiveness that are becoming so central to smart urban assemblages.

We have suggested a number of ways in which the urban has implications for energy, but we now turn to reflect on the consequences of smart urban power for the city. First we suggest that new forms of urban energy are deepening the sense in which the city is co-produced by a complex multiplicity of diverse actors, actants and processes. The partnerships and socio-technical accomplishments of smart grids considerably add to the degree of complexity in public–private arrangements that contribute to the constant re-establishment of the conditions necessary for everyday life in the city: power, heat, telecoms and so on. However, these projects are becoming more central, more essential to the fabric of urban governance, and to describe them as mere features of the contemporary city is unsatisfactory. Smart urban projects, among which smart power projects are some of the most significant and impactful, are (financially and geographically) sufficiently large and important that they can be thought of as becoming a dominant style of project-work in the city. While writers in the assemblage urbanism literature have emphasised the multiple political projects, modes of governance, practices and outcomes of cities as assemblages (McGuirk and Dowling, 2009), we suggest that within this picture of complexity and multiplicity there are some projects, some modes of government that are more central and more definitive of this new urbanism than others. We suggest that smart urbanism, and smart urban power as an important feature thereof, challenges the analytic neutrality of some assemblage thinking because of the sense in which a nervous system is being created as an end in itself, one which spans energy, heat, traffic, air quality and so on. This is a sensory apparatus in which, we acknowledge, there may be no single 'brain', no space or node of ultimate power or knowledge, but in which the process of constructing sensory capabilities and connections constitutes more than just one project among many. These are attempts to connect the otherwise feral forces of urban assemblage through a socio-technical but always political process which will continue to induce a power geometry through which winners and losers are created and, in different ways, revealed and obscured.

Notes

1   There are currently no commercially available, large-scale electricity storage technologies that can store electrical energy even at neighbourhood scale.
2   The US Department of Energy itself was awarded $36.7 billion under the 2009 American Recovery and Reinvestment Act to develop renewable generation and promote conservation and efficiency schemes across the country.

References

Biscoe, J. (2014) History of public supply in the UK [online]. Available at: www.engineering-timelines.com/how/electricity/electricity_07.asp [Accessed 22 April 2015].

Blok, A. (2012) Wandering around cities with ANTs: review of Ignacio Farías and Thomas Bender (eds) *Urban Assemblages: How Actor-Network Theory Changes Urban Studies*. *Science as Culture* 21 (2): 283–287.

Brown, L. (2012) World on the edge: how to prevent environmental and economic collapse. In S. Wheeler and T. Beatley (eds) *The Sustainable Urban Development Reader*. London: Routledge, pp. 205–214.

Butler, S. (2001) *UK Electricity Networks: The Nature of UK Electricity Transmission and Distribution Networks in an Intermittent Renewable and Embedded Electricity Generation Future*. London: Parliamentary Office of Science and Technology.

Covrig, C. F., Ardelean, M., Vasiljevska, J., Mengolini, A., Fulli, G., Amoiralis, E. *et al.* (2014) *Smart Grid Projects Outlook 2014*. Joint Research Centre Science and Policy Reports. Luxembourg: European Commission.

Crampton, J. (2007a) Key term: conduct of conduct [online]. Foucault Blog. Available at: https://foucaultblog.wordpress.com/2007/05/15/key-term-conduct-of-conduct/ [Accessed 22 April 2015].

Crampton, J. (2007b) What is the dispositif? Foucault Blog. Available at: https://foucaultblog.wordpress.com/2007/04/01/what-is-the-dispositif/ [Accessed 22 April 2015].

Deleuze, G. and Guattari, F. (2004) *A Thousand Plateaus: Capitalism and Schizophrenia*. London: Continuum.

Graham, S. (2000) Constructing premium network spaces: reflections on infrastructure networks and contemporary urban development. *International Journal of Urban and Regional Research* 24 (1): 183–200.

Graham, S. and Marvin, S. (2001) *Splintering Urbanism: Networked Infrastructures, Technological Mobilities and the Urban Condition*. London: Routledge.

Guy, S., Marvin, S. and Moss, T. (2001) *Urban Infrastructure in Transition: Networks, Buildings, Plans.* London: Earthscan Publication.

Foucault, M. (1980) The confession of the flesh. In *Power/Knowledge: Selected Interviews and Other Writings, 1972–1977*, ed. C. Gordon. New York: Pantheon Books.

Foucault, M. (1982) The subject and power. *Critical Inquiry* 8: 777–795.

Harvey, D. (1989a) *The Condition of Postmodernity: An Enquiry into the Origins of Cultural Change.* London: Wiley.

Harvey, D. (1989b) From managerialism to entrepreneurialism: the transformation in urban governance in late capitalism. *Geografiska Annaler – Series B* 71: 3–17.

Harvey, D. (2001) *Spaces of Capital: Towards a Critical Geography.* London: Routledge.

Hollands, R. (2008) Will the real smart city please stand up? Intelligent, progressive or entrepreneurial? *City* 12 (3): 303–320.

Hodson, M. and Marvin, S. (eds) (2014) *After Sustainable Cities?* London: Routledge.

Hubert, C. (n.d.) Smooth/striated [online]. *Writings.* Available at: www.christianhubert.com/writings/smooth_striated.html [Accessed 22 April 2015].

Hudson, R. (1989) Labour-market changes and new forms of work in old industrial regions: maybe flexibility for some but not flexible accumulation. *Environment and Planning D: Society and Space* 7 (1): 5–30.

Lewis, P. (2013) *Smart Grid 2013 Global Impact Report.* San Francisco: VAASA ETT.

McFarlane, C. (2011) Assemblage and critical urbanism. *City* 15 (2): 204–224.

McGuirk, P. and Dowling, R. (2009) Neoliberal privatisation? Remapping the public and the private in Sydney's masterplanned residential estates. *Political Geography* 28 (3): 174–185.

Massey, D. (1993) Power-geometry and a progressive sense of place. In J. Bird, B. Curtis, T. Putnam, G. Robertson and L. Tickner (eds) *Mapping the Futures: Local Cultures, Global Change.* London: Routledge.

New York Times (2009) An $80 billion start [online], 17 February. Available at: www.nytimes.com/2009/02/18/opinion/18wed1.html [Accessed 22 April 2015].

Ofgem (2011) *Smart Metering: What It Means for Britain's Homes.* London: Ofgem.

Patterson, W. (1999) *Transforming Electricity: The Coming Generation of Change.* London: Earthscan.

Pecan Street Inc. (2010) Working Group Recommendations [online]. Available at: www.pecanstreet.org/wordpress/wp-content/uploads/2011/08/Pecan_Final_Report_March_2010.pdf [Accessed 24 May 2015].

Shove, E., Pantzar, M. and Watson, M. (2012) *The Dynamics of Social Practice: Everyday Life and How It Changes.* London: Sage.

Smilor, R. W., Gibson, D. V. and Kozmetsky, G. (1989) Creating the technopolis: high-technology development in Austin, Texas. *Journal of Business Venturing* 4: 49–67.

South Tyneside Homes (2014) FAQs [online]. Available at: www.southtynesidehomes.org.uk/article/11365/FAQs [Accessed 22 April 2015].

Strengers, Y. (2013) *Smart Energy Technologies in Everyday Life.* New York: Palgrave Macmillan.

US Department of Energy (2014) Advanced metering infrastructure and customer systems. SmartGrid.Gov. Available at: www.smartgrid.gov/recovery_act/deployment_status/sdgp_ami_systems.html [Accessed 24 July 2015].

Verbong, G., Beemsterboer, S. and Sengers, F. (2013) Smart grids or smart users? Involving users in developing a low carbon electricity economy. *Energy Policy* 52: 117–125.

While, A., Jonas, A. and Gibbs, D. (2004) The environment and the entrepreneurial city: searching for the urban 'sustainability fix' in Manchester and Leeds. *International Journal of Urban and Regional Research* 28 (3): 549–569.

Wood, A. (1998) Making sense of urban entrepreneurialism. *Scottish Geographical Magazine* 114 (2): 120–123.

9
TEST BED AS URBAN EPISTEMOLOGY

Nerea Calvillo, Orit Halpern, Jesse LeCavalier and Wolfgang Pietsch

Introduction

There is a new utopian vision emerging in our present. Data-driven, designed to optimise information flows and the circulation of capital and constructed upon an infrastructure of pervasive computing and a fantasy of infinite bandwidth, the latest 'smart' or ubiquitous cities herald a new way to organise urban life. Whether lauded as a technology to manage human life under conditions of seeming apocalyptic security, environmental and financial threat, or critiqued as terminal threats to public and democratic life, these new urban infrastructures are remaking the reality and imaginary of what civic and urban space will look like. This chapter will investigate the ontology of these new spatial forms – these experimental test beds – for the future of human life.

Of all the many prototype spatial products currently marketed as 'smart' or 'ubiquitous' cities, the city of Songdo is perhaps the most infamous. It is one hour's drive southwest from Seoul and is being built from scratch on land claimed from the ocean (Figure 9.1). Songdo is one of a trio of cities that comprise the Incheon Free Economic Zone (IFEZ), a development initiated by the Korean government to attract foreign investment and residents. Songdo's distinguishing feature is the promise of its ubiquitous physical computing infrastructure. Marketed as a 'smart city', it is presented as an entire territory whose mandate is to produce interactive data fields that will, like the natural resources of another era, be mined for wealth and, similarly, will generate subsequent infrastructure for new forms of life.

The inquiry of the material effects that the mobilisation of the term 'smart' has had in this new city is the main objective of this chapter. Because of the lack of available information on Songdo's construction and development details we conducted a field trip to Songdo and Seoul in July 2012. We visited IFEZ on 4 July and spoke with Jongwon Kim (U-City business division, Incheon Free Economic Zone) and

FIGURE 9.1 Songdo International Business District under construction, Incheon Free Trade Zone, South Korea, 6 July 2012
Source: Jesse LeCavalier

Kyung-Sik Chae (Culture and Arts Division from Incheon Metropolitan). On 6 July, at the Tomorrow City welcome building we met Tony Kim (Director, IBSG, Cisco Systems Korea) and Gui-Nam Choi (Executive Services Sales, Cisco Systems Korea).[1] So this research is the result of personal interviews, corporate presentations, online brochures analysis, on-site personal experiences and direct observation.

The chapter is structured in two main sections. The first one looks at the role that data plays in the production of 'smartness'. The city is envisioned as a physical incarnation of an immense cloud of Big Data; its purpose and value are generated by speculation on how sensitive its infrastructure of sensors and cellular communication towers are, on how much data the city can generate and on how capable its high-bandwidth conduits are to circulate these data. IFEZ is like a holding company that is testing and refining a fleet of commodity cities that are imagined as mobile, plastic territories. They are simultaneously software, hardware, screens, algorithms and data (Crang and Graham, 2007), serving as interfaces and conduits into networks linked to other territories. These cities, like computational algorithms, are clearly defined and replicable: they are the protocols of a global infrastructure of information and economy. As a leading example, Songdo offers a unique vantage point from which to examine this global cultural and economic logic of large datasets.

The second section of the chapter unfolds the properties of *test-bed urbanism*, the name we have given to the distinguishing modes of operation we saw at stake. Like Enlightenment utopias such as the *New Atlantis*, governed by a specific scientific practice of empirical experimentation, Songdo is also touted as an ideal site for new forms of experimental practice. For example, Songdo has been referred to as 'the experimental prototype community of tomorrow' (Lindsay and Kasarda, 2011: 4; Lindsay, 2010: 90). Furthermore, implicit in our discussions with Cisco, Songdo is understood as a model that can be bought, replicated and deployed (Choi, personal communication, 6 July 2012). However, unlike the older forms of Enlightenment science, this experiment does not subscribe to the same rules. We argue that Songdo is reflective of a new form of epistemology that is not concerned with documenting facts in the world, mapping spaces or making representative models, but rather with creating models that *are* territories. Performative, inductive and statistical, the forms of experimentation enacted in this space transform territory, population, truth and risk with implications for representative government, subjectivity and urban form. These features of what we are calling *test-bed urbanism* are increasingly evident globally, both in new 'smart city' projects (Hollands, 2008; Ho Lee *et al.*, 2008) but also within the discourse of urbanism more generally (Augé, 1995; Koolhaas and Mau, 1995; Elden, 2007). We argue that this test-bed urbanism is a form of administration and a redistricting of bodies and information into new global configurations that are increasingly affecting all our lives and therefore demand explication. We do this by first examining the operations 'on the ground' at Songdo in order to assemble initial evidence in support of our concluding speculative claims concerning the epistemology of the test-bed city.

To conclude we interrogate the possibilities of new cities marketed as 'smart' in proposing new urban futures. Songdo, unlike other ideal cities whether built or unbuilt, has no perfect whole and is thus both literally and conceptually incomplete. However, it remains utopian in the sense that it aspires to achieve new forms of life, even in a perpetually provisional version (Gordin *et al.*, 2010). This city is a rehearsal of our future and an archive of our past. The purpose of this chapter is to excavate this wishful thinking and to examine the tense relationship between performance and aspiration. Like all utopias, Songdo is also a 'heterotopia', a space that can tell us about our world, make us conscious about the choices – aesthetic, architectural, designed and technical – that we are making and still have to make. Most importantly, these mirror worlds – dystopian, ugly, banal, beautiful – provide us with visions of alternative realities and portents of events we might seek to avoid (Foucault, 1986). Such spaces make us realise that what we assume to be natural – the desire, for example, for a 'smart' planet – is contested, situated and historically specific. In Songdo the present is not known and the future is not already here.

Arriving on new shores

The phrase 'smart city' can be feasibly applied to a large number of diverse international projects that range from the updating of telecommunications infrastructure to the construction of entirely new, planned cities. Hollands points out that already in 1997, the World Forum on Smart Cities suggested that there would be 50,000 cities and towns working on smart initiatives by 2010 (Hollands, 2008: 304). The common denominator in most of these projects is an investment in digital infrastructure and a belief in the possibility of 'improving life through technology'.[2] These projects operate at a range of scales: from the development of a 225,000-person 'smart valley', as in Portugal's Living PlanIT project, to the 2.5 million people projected to eventually settle in Songdo, one of the largest smart city projects to date. Gale International, a Boston-based real estate company, is developing Songdo with the support of the Incheon municipal government. The new city is part of the larger Incheon Free Economic Zone (IFEZ), a new juridical designation that also includes Cheongna and Yeongjong. Songdo is connected to the Korean mainland by electronically monitored bridges and is strategically located close to the Incheon International Airport. IFEZ actively courts foreign investment and labour through significant tax incentives, logistics capabilities, leisure opportunities and the promise to be one of the world's 'smartest' urban regions (Jongwon Kim, personal communication, 4 July 2012). In pursuit of this goal at Songdo, Gale enlisted San-Jose-based international networking company Cisco to develop many of the smart technologies and services and attendant infrastructure (Cortese, 2007; Lindsay, 2010).

Famous for building routers and infrastructure for networks, Cisco now aspires to become a management-consulting corporation, with the expertise to build the informational infrastructure for cities of millions of people overnight. Their main concern is increasing the demand, therefore, for bandwidth. They hope to produce what their competitor IBM has labelled a 'smart' planet. IBM has an entire new management-consulting service branded around 'smart planet' services.[3] Cisco is planning to also retrofit itself into a more consulting-service-orientated company rather than only or mainly sell hardware like routers for digital infrastructure. As part of its rebranding and diversification efforts it initiated its Smart+Connected Communities (S+CC) programme in 2009, a 'global initiative using the network as the platform to transform physical communities to connected communities run on networked information to enable economic, social and environmental sustainability' (quoted in Cisco, 2011: 1). In building these cities, Cisco's role is largely as the management consulting and strategy firm for high-tech services. The conduits, routing systems, sensors, telecom towers and other hard portions of the infrastructure are built by telecom companies Cisco is partnered with, in this case Korea Telecom (KT), and the buildings are built by GALE International, a New York City-based developer, also going by the name of KPA, that created the master urban plan (Kim and Choi, personal communication, 6 July 2012). 'Smartness' is what ties them all together.

At the centre of Songdo's marketing materials and technical discourse lies a fantasised transformation in the management of life – human and machine – in terms of increased access to information and decreased consumption of resources. The developers, financiers and media boosters of this city argue for a speculative space ahead of its time that operates at the synaptic level of its inhabitants. So for example, there is a great predilection for implanting LED screens everywhere in the urban space[4] and pushing video conferencing to be integrated with other bio- and labour-monitoring devices. Video demands high bandwidth above and beyond all other media formats. Cisco's strategic planners envision a totalising sensory environment in which human actions and reactions, from eye movements to body movements, can be traced, tracked and responded to in the name of consumer satisfaction and work efficiency. Of those we spoke with at IFEZ and Songdo, most agreed that no one could really define these terms, but they were useful as a goal and ideal. It is worth noting that whatever these terms may denote, they are always extendable, as, incidentally, are the concepts of 'intelligence' and 'smartness'. Every wall, room and space is a potential conduit to a meeting, a separate building, a remote lab or a distant hospital. For example: marketing videos showed the roll-out of tele-medicine applications, which required, as some engineers suggested, transforming the laws of South Korea to allow the construction of medical-grade networks to allow genetic and other data to flow from labs in the home to medical sites in the proliferation of home health-care services (Kim, personal communication, 4 July 2012; Kim, personal communication, 6 July 2012). The developers thus envision an interface-filled life propelled and organised by a new currency of human attention at its very nervous, or even molecular level.

This manipulation of synapses – both those of the city and those of its inhabitants – is closely linked to a second discussion about the resulting outgrowth of sustainability as a guiding principle. Consequently, the two discourses that shape the speculation around Songdo and offer insight into the infrastructural logic of digital media are those of, on one hand, preparedness for possible ecological disaster (i.e. sustainability) and, on the other, a ravenous and expanding capacity for attentive manipulation and management of information, resources and inhabitants (i.e. bandwidth). The latter stretches into the very minds of the inhabitants of the city, who, incidentally, are also increasingly imagined as components in an urban-cum-global network. From this infinite set of biological, machinic, demographic and environmental data, a fantasy of self-regulating and self-propagating systems emerges. This is a city that may not be fantasised as conscious, but certainly as something capable of intelligence and cognition by way of modulating and measuring the affective states and senses of its many inhabitants – human, machine, or otherwise, a notion reinforced even by the adoption of the term 'ubiquitous' to describe the city's impulsive and subconscious operations (Weiser, 1991; Ho Lee *et al.*, 2008).

Data, an urban resource

In spite of its claims of greenness and smartness, Songdo remains challenged by its profit motives and is breeding new entrepreneurial opportunities and institutional structures as a result. In Songdo, Cisco and IFEZ are experimenting with the

FIGURE 9.2 Seoul urban screens, South Korea, 8 July 2012
Source: Nerea Calvillo

construction of a joint venture between public and private companies by which the private sector invests in new data infrastructure with the promise of accessing and using public data in return. The public–private cooperation company (PPCC) provides a combination of services, but the main characteristic is that users pay the PPCC directly for elective services, rather than expecting them to be delivered. Cisco hopes that this will make for a more profitable and more effective way of developing new technology. At the same time, in the case of Songdo, the municipal government of Incheon hopes to use the partnership to more effectively finance the services it provides.

In either case, both Cisco and IFEZ are looking for new sources of revenue and hope to 'monetise' the attentive capacity of Songdo's inhabitants. Their hope is to use this latent reserve of data gathered on users to produce services that can be paid for through advertising in multiple-scale devices (Figure 9.2), electronic education, physical treatment, home tele-medicine or any number of other speculative services vying for a share of this new market. For Cisco – like Facebook, Google and other companies that attempt to link user behaviour at the interface with consumer behaviour in order to monetise their vast datasets – data is the currency of this new realm, a realm envisioned as an interface for inserting and extending the sensorium. Songdo is, thus, a parody of the fantastical perpetual motion machine of the nineteenth century: a system that theoretically continues to produce wealth-without-end through the construction of huge conduits for bandwidth and of vast quantities of environmental sensors, all focused on the monitoring and indexing of its inhabitants' on-line and off-line behaviours. These self-referential and self-generating properties make Songdo, perhaps unsurprisingly, mimetic of the logics of the very financial systems that have conceived and sponsored this 'product.'

Automated infrastructures

The individual elements that comprise this sensing and data-recording system are inert devices designed to absorb input and direct it to a processing centre that aggregates it and analyses it. Sensing devices will be ubiquitous features of the city and will be active in both domestic and public spheres. For the latter, Songdo will use an integrated sensing element called a 'smart pole' that provides light, sound and navigation information (Figure 9.3). It is also equipped with a CCTV camera and emergency broadcasting hardware (Cisco, 2012: 15). The smart pole can also receive input either though a callbox with an emergency button or through an Internet terminal. These poles will be installed at regular intervals throughout the city and will be capable of both responding to and producing a range of environmental conditions. The input collected by any given smart pole is directed to the 'integrated operating center' (IOC) that, in turn, analyses the data and sends back commands to the source.

A control room is one of the dominant features of an IOC and is an interface that conveys to human monitors whatever the digitised urban environment unveils. As such, its power is feeble, its knowledge is limited and its vision is blurred. Moreover, though it is prepared to handle emergencies and clad in the aesthetics of Cold War preparedness, the control room is primarily a site of maintenance (Figure 9.4). Thus, the monitors' role is one of management and not necessarily one of protection. Monitors survey the changing array of images in search of any perceived disruptions to the system.

In some cases, human monitors bear witness to events unfolding within a camera's cone of vision but, more often, the large number of recording devices makes keeping track of all these data impossible. Thus, the task of extracting relevant information is increasingly handled automatically. The often-used but still revealing phrase 'data mining' reinforces the implicit understanding of these practices. Songdo's sensors act as a fleet of interconnected agents that track the behaviour of human and non-human inhabitants of the city, turn their recorded activities into data to be 'mined' and eventually extract profitable information from this ethereal ore (Lohr, 2012).

At the domestic level, it is assumed that every wall, every mirror and every surface can become an interface that offers users everything from on-demand data and weather reports to home medical monitoring. Demonstration videos show domestic spaces fitted out with home genetic-testing kits, blood-work labs and heart-monitoring stations, all to ensure the health of the residents of these luxury

FIGURE 9.3 Smart pole detail
Source: Nerea Calvillo

FIGURE 9.4 Demonstration control room, City of Tomorrow, Songdo
Source: Jesse LeCavalier

FIGURE 9.5 Sensor data can occasionally reach conclusions that are self-evident
Source: Nerea Calvillo

high-rises. And yet this does not only have technical implications. The current legal system in Songdo is being lobbied to enact changes in privacy laws that would allow the transfer of medical data outside of the hospital in order for Cisco to roll out its medical tele-presences services.

But just what do we see when we look at this universe of doubly communicative surfaces (in that they both send and receive)? In the promotional material from Songdo, these screens are interfaces that help viewers make better decisions to save time, money, energy or anguish. Through this promise of omniscience and omnipresence viewers/users/consumers can exceed their human limitations thanks to the automated collection and analysis of data that is suddenly easy to access. But these interfaces work *on* us as much as *for* us. The bilateralism of the interface informs users but also makes them *informers* – i.e. it works to optimise viewers and the network in which they operate. Because their habits and desires create a map of future habits, supply and demand will eventually merge.

In the meantime, the monetisation of human attention continues as companies like Cisco intensify their research into ways to deploy and capture information. The limiting factor in increasing rates of data transmission is the capacity of the hardware and of the network to transmit information. Cisco's turn to urban development and to the production of smart city models and prototypes is an exercise in creating markets for the very hardware on which the company was founded. In fact it is data, as a conceptual entity, that drives the ambitions of the city.[5] The dream of data ubiquity described by Mitchell in 1996, and still present, reinforces the concept of extendable value that emerges through the logic of constantly testing new functions, new products and new applications. Songdo is driven by a fantasy that by translating everything into data, the whole city could be managed as if it were a continuous and apprehensible system. This capacity for ubiquitous sensing is instrumental to a new manner of relating to the environment and, in these new cities, materialises in an architecture of sensing devices, communications towers and fibre-optic cables embedded throughout the built space (Figure 9.5).

Performative normality

In this desire to 'data-fy' the environment through the extreme expansion of bandwidth, any inhabitant, human or non-human, is considered measurable in different ways and all inhabitants are treated as equal entities in terms of the data they can provide. For example, the movement of people in the city, water velocity in a sewer, personal energy consumption, shopping habits, browsing patterns or particulate matter suspended in the atmosphere, to name just a few, all feasibly contribute to the construction of a more responsive urban environment (Kim, personal communication, 6 July 2012). Because of the multiple but equally valid sources for these data, it no longer makes sense to address them in terms of qualitative or quantitative, subjective or objective. Rather, their definition is no longer about truth to nature or to an external world but is expanded to incorporate the emotional, the affective and what was previously non-formal knowledge. Even the traces of transactions known as 'shadow data', including phone conversations, credit cards or movements through the city, are stored and made ready to be recycled for other purposes. However, if until now the choices made about what to collect and why have had histories of controversy and resistance, in the test-bed city, the attitude is to collect indiscriminately and to accumulate by default.

The law of probabilities governs Songdo and, thus, its mining of past data is done in the name of the future. While those responsible for the development of Songdo need *users* before they can manipulate user data, they are nonetheless proceeding with the implementation of the necessary sensing apparatus, even if its ultimate end use remains unclear. However, even if the precise future of the data cannot be predicted, it will most certainly be analysed and cross-referenced automatically. The promise of such number-crunching is that we will learn previously unknown things about ourselves based on an idea that the collective behaviour of a city can be compiled and analysed by machines in order to reveal profound trends in our social behaviour. However, this approach could also easily, and perhaps dangerously, produce a number of false correlations (Spade, 2011). A likely outcome of these urban-scale calculations is a modification of concepts of normality. Rather than measuring similarities in terms of their deviation from an established norm,

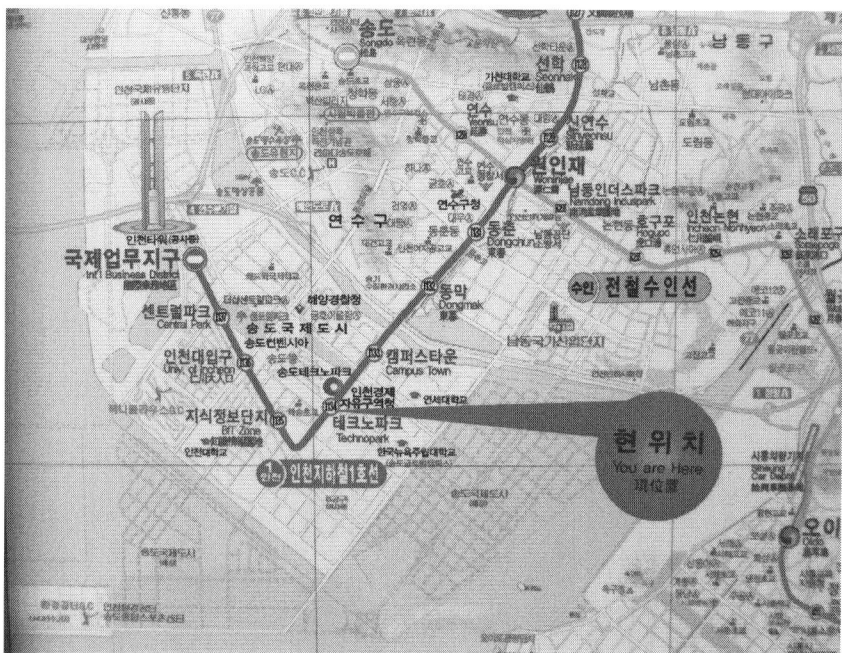

FIGURE 9.6 Songdo is constructed by dredging sand from the ocean flow and consolidating it to create a new landform
Source: Jesse LeCavalier

producing what Ian Hacking (1990) has called 'the taming of chance', or by an overlap of different normal curves producing an 'interplay of different normalities' (Amoore, 2009: 55), the prediction models in the test-bed city are non-normative. Instead 'normality' is constantly constituted as it emerges from the summated actions of daily life.

This is not to say that all reference points are removed, only that normative conditions index performance rather than ideal forms. Consider the actual production of territory in Songdo as an example. In ideal cities of the past – a lineage in which we can place Songdo – city makers were preoccupied with the pursuit of perfect geometric conditions (Rosenau, 1983). The form of Songdo, by contrast, is conditioned by performance requirements and materialises the countless and competing protocols to which the city must submit in pursuit of global competitiveness (Easterling, 2004). Such requirements include things like sand-dredging techniques to create a new landform (Figure 9.6), channel depth for container ships and demographic analysis that predicts housing demand or communications connectivity. So while Songdo is an accommodating vessel, it is a vessel nonetheless; a medium through which other media pass, including the increasingly datafied modules of international logistics systems. The idea of a protocol-driven – or algorithmic – expansion of space can expand beyond computational infrastructure to

FIGURE 9.7 Model of Songdo in the City of Tomorrow showroom
Source: Jesse LeCavalier

encompass the very logic by which territory, liquid and informational, comes into being. Freed from the model of an ideal form, the city can be enhanced without limit (Figure 9.7).

The test bed is an epistemology

Songdo is, arguably, the most extreme instantiation of a far more prevalent and genuinely ubiquitous faith in the place of big data and interactive feedback to monitor and sustain daily life. The technologies tested in Songdo are beta versions of similar systems put in place by many cities, and rolled out regularly by high-tech and telecommunication corporations to service the now naturalised faith that human beings require ever more information and bandwidth for social life. If many technologists and planners view these massive greenfield spatial products as banal and uninspired, it is at the cost of failing to realise that the array of conduits and cell-phone towers of IFEZ are practice spaces for corporations to perfect the design of data collection and management infrastructures for any network – urban or otherwise.

If this is true, then what type of territory is Songdo? What can these test beds for urban construction help us see about this broader logic of testing and big data as forms of governmentality – those techniques of measurement, regulation and monitoring – that organise space and manage life (Foucault, 2009)? What types of knowing and acting are facilitated by way of test beds, and what makes them specific to our contemporary condition?

The test bed is an engine, not an image

A 'test bed' is not an experiment as conventionally conceived in the ideals of science. The term does not denote the unearthing of a truth about the world. The phrase 'test bed' emerges in the engineering literature to describe a controlled and often isolated development environment in which to test the operability of new technologies, processes or theories for large systems. Test beds can include practices such as beta-testing software, testing control systems in manufacturing, stress testing in financial regulation and so forth. In keeping with the logic of the present, Songdo is a test bed for a form of urban life that is itself the product – much like the financial instruments that Donald MacKenzie describes as 'engines, not cameras' as he argues that financial equations produced to model markets also produce markets in the equations' very circulation and use (MacKenzie, 2006). The same is true here. This city is an engine for urban growth even as it purportedly tests the capacities of its own infrastructure. In essence, it is an experiment that cannot end, because every limit becomes a new challenge, a new frontier to develop toward an ever-extendable horizon.

This extension relies on a certain refusal of intelligence, objectivity and representation. Like the swarms, insect colonies and chaotic systems that today are the sources of inspiration for computer scientists in producing technology and network architecture, Songdo is also built from a constant logic of dumbness, dispersal and speculation. Self-organisation is its dominant rationality. There is no image, because this form of 'smartness' is not imagined as conscious, liberal or objective. Rather intelligence here is linked to performativity. Across the computational sciences, ideas of intelligence have ceased to be about linguistics or representation, and turned instead to building smart systems from the complex interactions of many basic actions (Terranova, 2004: 36–7). In Songdo what is imagined is a generalised smartness that emerges from the sensor networks distributed all over the 'body' of the city. No single sensor has the full picture of the environment, but as a collective they are able to act as if they did. If challenged, sensors can regroup as a network of statistical agents, each can react to the information they are provided with and each can adjust its actions accordingly.

More importantly, this approach to environment, planning, citizens, subjects and intelligence marks a turn against the faith in liberal subjectivity, denigrates the place of older political processes in decision-making over infrastructure as a site of activity and operates at a level far beneath consciousness. This is a dream of a world that operates through networked nerves that hook the sentiments, feelings and movements of live bodies into larger circuits of capital and technology, without (at least in the aspirations of the engineers) passing through the filter of representation (Figure 9.8). In the test bed the contours of the city are never stable, always modifiable and the

FIGURE 9.8 Communal garden inside a park
Source: Nerea Calvillo

territory is extendable. The engineers are almost dismissive of the form of the city as it is envisioned as governable by response to itself. As such, unlike other utopias, the habitable form of the city and its machine function are neither ideal nor closely linked. This is both a blessing and a curse. Without ideal or aspirational images, the future becomes blurry, but it also can become more open and extendable. The loss of centred subjects, the proliferation of surveillance and sensing, also comes with the decentralisation of control and the proliferation of sites of action.

The test bed calculates uncertainty, not risk

This lack of an ideal is not therefore merely the concern of connoisseurs of form. Rather it is the mark of a method for speculation. This utopia of the version founds itself on a particular statistical and probabilistic logic based in feedback. Here the distinction Arjun Appadurai makes between risk and uncertainty might be useful. For planners there is a great deal of uncertainty in the outcomes of their experiments, but what has increased is a faith in 'techniques of calculability' (Appadurai, 2011: 528). This faith, and it takes on mystical and magical undertones, embeds itself in the sensory environments and business models that largely bank on fantasised datasets that can be used to direct every aspect of life – from the medical status of bodies to education to entertainment.

This data is considered valuable because of an armament of new techniques for excavating results without ever knowing the objective endpoint or baseline. Its lack of a definite endpoint or image of the city is indicative of the new calculations that take into account uncertainty, but cannot think from within their internal logic about absolute failure, loss or termination – if only because there is no base point from which to start. Repeatedly, both government officials and engineers argued that this was a prototype whose failings would be fodder for the next generation of cities.

The capacity to envision such architectures of speculation and computability emerges with global transformations in attitudes to networks and computation. These transformations include a shift from a deterministic to a probabilistic mode of thinking in artificial intelligence, communication theories, social sciences and economics, beginning after the Second World War, and a concomitant interest in rethinking intelligence or 'smartness' as distributed, performative and not representative or conscious.[6] If digital infrastructures determine the city of the future, then their characteristics will have a hidden, largely imperceptible but highly formative influence on the way we live, and more importantly *will* live, our urban lives.

Songdo is, therefore, an enormous gamble for both urbanists and its developers. It is, in fact, the largest private real-estate project on earth (Arthur, 2012). In order to understand how such capacities to make such investments occur, its important to understand how Songdo is one material manifestation of a far broader faith in new forms of statistical calculation to mitigate and disperse risks into a new term, 'uncertainty'. It is helpful to recall that the analysis of big datasets bases itself in methods such as Bayesian filtering and statistics. What marks these methods is that they deal with quantities that cannot be discretely measured or stably represented, which is to say they lack an 'image' in our parlance. Frank Knight in 1921, in a critical treatise on economics,[7] articulated this distinction:

> But uncertainty must be taken in a sense radically distinct from the familiar notion of Risk, from which it has never been properly separated. The term 'risk,' as loosely used in everyday speech and in economic discussion, really covers two things which, functionally at least, in their causal relations to the phenomena of economic organisation, are categorically different ....The essential fact is that 'risk' means in some cases a quantity susceptible of measurement, while at other times it is something distinctly not of this character; and there are far-reaching and crucial differences in the bearings of the phenomenon depending on which of the two is really present and operating. There are other ambiguities in the term 'risk' as well, which will be pointed out; but this is the most important. It will appear that a measurable uncertainty, or 'risk' proper, as we shall use the term, is so far different from an unmeasurable one that it is not in effect an uncertainty at all. We shall accordingly restrict the term 'uncertainty' to cases of the non-quantitative type. It is this 'true' uncertainty, and not risk, as has been argued, which forms the basis of a valid theory of profit and accounts for the divergence between actual and theoretical competition.
>
> *(Knight, 1921: online)*

Bayesian inference is one example of numerous methods deployed to operate on precisely such unmeasurable, subjective and qualitative endpoints. Such methods mark a turn to inductive reasoning, subjective perspective (there are no stable truth claims) and the abandonment of stable baselines or norms; a turn that finds itself incarnated in such ideas as 'data-driven' science, marketing and strategy. To be 'data-driven' is to start without ideal or hypothesis, to assume no stable baselines and to always modify your conclusion. Risk is replaced with constant testing and versioning.

If the test-bed world of big data fields is a probabilistic one where few things are certain and most are only probable, the test-bed world is statistical. The analytic methods of big data, including Bayesian statistics, surmount this qualitative/quantitative divide, however, by never defining the problem or hypothesis ahead of time. The new modes of calculation deploy technologies that are algorithmic and Boolean in manners that support the management of non-measurable and indeterminate outcomes. Developers, financiers, management and high-tech companies all work inductively, simply from the behaviour of the system rather than from deductive theories that must be tested. What is less clear is what are the conditions and limits for what can be sensed and assimilated.

As past utopias described a dream world of empirical study and deductive reason, an emerging utopian dream is to bridge a world of discrete things – objects, algorithms, logic definitions – with a world of free-flowing infinity. The Gothic cathedral and the spatial dream of infinity will come into the network and will cover the globe and beyond. So infinite is this dream that we no longer can even visualise it. The loss of the ideal image of space is replaced with an ideal of the perfect methodology.

This is the law of the test bed. In the world of the test bed there must always be more data and there is never a stable endpoint. Time itself is broken. There are no events in the test bed. Disasters, like accidents, ecological degradation and medical emergencies, are regularly detailed but are not terminal or measurable events. Instead, they are ongoing processes that can be manipulated and managed through constant feedback. This is a city built to anticipate the uncertain and to accommodate it through the decentralised nature of networked infrastructure that has itself automated emergence and change as regular and manageable processes.

The test bed thus transforms time, change and events into uncertainties and trials. This is a new form of administration that lacks norms, frequency distributions and the statistical apparatus of older demographic, state and economic thinking in the name of a new epistemology of infinity, non-normativity and speculation. While this is potentially liberating, the loss of a norm also undermines moral grounds for action. Political decision-making is constantly deferred and managed technocratically. But these instabilities and emergences also produce strange actions in the network with often surprising results. As critiques of modernity have long noted, the concept of a rational, measurable endpoint – utopia – has led to catastrophe. Having lost this vision of the ideal city we must learn to embrace the manipulation of probabilities without idealising the concept of emergence. History and politics need to be rethought in terms of probabilities, densities, distributions and performances.

The test bed is about populations and territories, not individuals in space

We opened this chapter with *New Atlantis*, a historical reference important to historians of science and politics. Francis Bacon illuminates, at the start of the scientific and political revolutions of the seventeenth century, a transformation in the ways that society might be governed. According to Michel Foucault, Bacon made visible the historical emergence of an idea that the terms reason, rationality and ratio are fundamentally linked. For Foucault, only in the late sixteenth and seventeenth century did the discovery of ideal laws of nature (ratios) get linked to the reason of government (Foucault, 2009: 277). This correlation between knowledge and power transformed space, as well. The state is not just the area administered by a sovereign, but a space that has qualities that can be measured, rationalised and experimented on. Space became *territory*, an area defined not merely by physical geography but by ratios. Territory, thus, is not an area governed by a sovereign but an area concerned with the security (the sustainability, maintenance and qualities) of a population. As geographer Stuart Elden states, 'Just as the people become understood as both discrete individuals and their aggregated whole, the land they inhabit is also something that is understood in terms of its geometric, rational properties, or "qualities"' (Elden, 2007: 278). This is another way of saying that the modes of discovering the rules or 'ratios' that govern a space become the role of government. This is not just about counting, but speculating: statistics and geometry are linked in seeking to unearth recurring patterns that are scalable in time and space, and can allow speculation on the future forms of the population, their actions and the shape of the space they will occupy.[8]

What marks the new territories of ubiquitous computing and test-bed experiments is that these 'qualities' can be redistributed and partitioned with growing plasticity, and their populations are not comprised of individual bodies, but of literal partitions of attention and nervous energy that can be grouped into different 'wholes' that are unstable and, like the territories they occupy, mobilised, circulated and speculated upon.

In this newest state of government the population and the territory have become the same – which is to say the territory being sold is based on attentive consumption and the monitoring of that consumption by groups of individuals. These individuals can be anywhere in the world; the territory is plastic and it does not need to be occupied (Figure 9.9); it is composed of all those who download medical assistance, yoga lessons or language lessons, for example. It is also composed of all the other spaces where similar infrastructures are being laid down, and cities built on the premise of ubiquitous data wealth. Songdo therefore is an experiment in administration, territory production and finance.

Songdo is not testing which kind of data is measured or manipulated, but *how* it can be managed so that the whole process can be exported elsewhere. This is the territory of management and algorithmic processes. All the agents – individually and as teams – are packaging their knowledge, expertise and services, so they can

FIGURE 9.9 Underground access in Songdo
Source: Nerea Calvillo

sell it to other cities. For example, Cisco will be importing the entire master plan of Songdo to the city of Guayaquil, Ecuador, thus literalising the idea of an exportable city. It is this exportability that matters in the test bed for that is what can be implemented in future phases of development. But this model presents Cisco, once again, with the fundamental problem of the lack of compatibility or common formats that can ensure the easy exchange of data. For this reason the Korean government, in partnership with Cisco, is now leading an international project which tries to develop the protocols of the smart city, having Internet protocols as a reference, by which data can be exported, recombined, homogenised and linked.

Conclusion

Examining cities that are built out of nothing affords a direct, if uncomfortable, confrontation with some of the assumptions about what cities are, or at least what they are thought to be. More disquieting, we must contemplate our dream-images of the future.

Looking at this seeming simultaneous wasteland and wild frontier of digital speculation, we see vast towers built as some parody of Le Corbusier's 'City of Tomorrow' (Figure 9.10). We should ask ourselves why – if digital infrastructures could generate any form – we are so limited by older imaginaries of vision and

FIGURE 9.10 Towers around 'Central Park', at the core of Songdo
Source: Orit Halpern

structure. Various authors have discussed the implications of computation in the definition of space and urban form. 'Space', writes Tiziana Terranova, 'does not really need computers to be informational even as computers make us aware of the informational dimension …. An informational space is inherently immersive, excessive and dynamic: one cannot simply observe it, but becomes almost unwittingly overpowered by it' (Terranova, 2004: 37). As Ash Amin has written – referencing the work of Mbembe (2004) – on the logics of urban forms, since the urban unconscious is 'composed of a city's material infrastructure as well as a city's "aesthetics of surfaces and quantities," it can be thought of as a field of affective excess that is able to "hypnotise, overexcite or paralyse the senses"' (Amin, 2010: 6). These accounts describe a situation in which, like early experiments in utopian form, urban inhabitants wish for a total vision of the environment, rather than embracing our machinic attributes and what Donna Haraway has called a 'partial perspective' (Haraway, 1991). What all these theorists gesture toward is our inability to understand the networks in which we are enmeshed. We should not be so sure of the present and past verdict on our networks. For if the return to modernist utopian visions in planning seems horrible, no less horrid is a nostalgic lament for the beauty of human community before the time when machines, also, were bequeathed sentience. Our networks are often more lively than we can predict. If the test bed is our new epistemology we should err towards the incalculability of uncertainty rather than the measurable logics of risk.

In returning to these older fantasies of measure and objectivity – from modernist utopias and computation – we threaten our own future. Perhaps this is the problem: in the logic of the test bed, past data is always used to produce the future. But when prediction collapses into production, we lose any possibilities of emergence, of change or of dynamic life. We often assume we understand complex systems and our machines, even if we do not or cannot. Perhaps, then, we can embrace the epistemology of the test bed and begin to design with less authority and greater interest in the space of society and culture that is produced in the interstices between what is human, machine and more than human. Perhaps it is out of the magic of our humming networks and the complexity of our mute black boxes that new ideas about design will emerge.

Even as we write, the ontology of cities is changing rapidly and dramatically. Old categories that have served well for centuries are becoming obsolete, like urban versus rural, infrastructure versus built visible structures, home versus work. The topographies, the spatio-temporal relations among citizens and with respect to the places they inhabit undergo profound changes as well. Distances diminish as the costs and speed of transferring goods and information decrease. And the various layers of city networks, the material and digital skeletons of the city, extend, branch and interweave, causing traditional boundaries to vanish, between different infrastructures like information and energy, between the service and the serviced, between the citizen and the city.

The new ontologies that arise are digital and algorithmic; they are often elusively abstract like data densities, clouds, statistical risks or visibilities, and sometimes oppressively concrete like ubiquitous cameras, secretive control rooms or windowless data centres serving the needs of the machines. Novel actors populate newly emerging city spaces, displacing older inhabitants of the cityscape. The data centre ousts the office space, cameras the watchman, sensors the maintenance technician and so forth. The city ceases to be modular, with unforeseen consequences regarding emergencies and control. Data is the main category that drives the transformation of modern city life and rearranges the hierarchies and connections of objects and people.

The impact of these changes is still mostly felt around the edges of our most massive urban spaces, in places like the peripheries of rampant greenfield cities in China or India that need to accommodate an ever increasing influx of people. Or serve as the latest gated communities. Or in context-free and replicable environments, like Songdo. However, as a city still being built the effects in daily life of its 'smartness' are yet to be seen. What has been demonstrated is that the notion of 'smartness' has mobilised a process of experimentation where global corporations have had to deal with local governments, landscapes and investors to test business models, infrastructural deployments or marketing strategies. So 'smart' has produced in Songdo a test-bed urbanism which is not unique to this Korean city, but is being enacted around the world in cities like Masdar, just with other companies, governments and territories. What makes Songdo a fascinating case study is that all its properties are more visible, because it was the first smart city built from scratch, it has been constructed on artificial land and it is big.

But projects like Songdo are also test beds for the larger revolution of urban life that will increasingly infiltrate even the most settled of urban environments. Such test-bed urbanism accompanies the digital revolution of an urban life trying to assess the impact of the algorithmic on tomorrow's citizens. Never before in history have cities been subjected in such scale to the technocratic visions and trials of a few anonymised global companies. But never before have there also been so many new agents and agencies – human, machine and other – networked in new arrangements and intelligences. These experiments will concern us long before their outcome will be clear. But, like the many speculators and corporations trying to cash in on these developments, only to be frustrated in their financial ambitions, the present is rather cloudy and the future often unpredictable.

Acknowledgements

The authors would like to thank the Institute of Public Knowledge at NYU, the BMW Foundation and the Herbert Quandt Foundation for sponsoring and organising the Poiesis Fellowship, under whose auspices this research was conducted.

Notes

1   We are very grateful to all of them for their help and generosity.
2   For more information, see 'PlanIT Valley – the benchmark for future cities and sustainable urban communities'. http://living-planit.com/planit_valley.htm.
3   www.ibm.com/smarterplanet/us/en/?ca=v_smarterplanet (accessed 7 August 2012).
4   On the implications of urban screens in the constitution of public space see Calvillo, 2012.
5   Here it is useful to contemplate what form of data we are discussing. When we talk about data we borrow Kitchin and Dodge's distinction between data and 'capta': with respect to a person, data is everything that is possible to know about a person, and capta is what is selectively captured through measurement" (Kitchin and Dodge, 2011: 5).
6   For work on time, epistemology and the nature of probability in digital media see Halpern, 2005.
7   Despite being written in 1921, this essay took prominence only after the Second World War, with the rise of Chicago School Economics and neo-classical economics. Milton Friedman, for example, was a student of Knight's Library of Economics and Liberty.
8   For another in-depth discussion of how numeracy, measurement and logistics are transforming territories see, for example, LeCavalier, 2010.

References

Amin, A. (2010) Cities and the ethic of care for the stranger [online]. Joseph Rowntree Foundation/University of York Annual Lecture. Available at: www.jrf.org.uk.sites/files/jrf/cities-and-the-stranger-summary.pdf [Accessed 24 May 2015].
Amin, A. and Cohendet, P. (2004) *Architectures of Knowledge: Firms, Capabilities, and Communities.* Oxford and New York: Oxford University Press.
Amoore, L. (2009) Algorithmic war: everyday geographies of the war on terror. *Antipode* 41 (1): 49–69.

Appadurai, A. (2011) The ghost in the financial machine. *Public Culture* 23 (3): 517–540.
Arthur, C. (2012) The thinking city. *BBC Science. Technology. Future FOCUS* [January] (237): 55–58.
Augé, M. (1995) *Non-Places: Introduction to an Anthropology of Supermodernity*. London: Verso.
Calvillo, N. (2012) Media structures, prototypes and collective prosthesis. In S. Pop and U. Stalder (eds) *Urban Media Cultures*. Ludwigsburg: Avedition GmbH, pp. 84–87.
Cisco (2011) Cisco Smart+Connected communities: media backgrounder [online]. Available at: gov.ns.ca/govt/pubs/Cisco-SCC-Media-Backgrounder.pdf [Accessed July 2015].
Cisco (2012) Smart City Technology and Advisory Services. PowerPoint slide presentation by Tony Kim and Gui Nam Choi, 4 July.
Cortese, A. (2007) An Asian hub in the making [online]. *New York Times*, 30 December. Available at: www.nytimes.com/2007/12/30/realestate/commercial/30sqft.html?pagewanted=all [Accessed 24 April 2013].
Crang, M. and Graham, S. (2007) Sentient cities: ambient intelligence and the politics of urban space. *Information, Communication, Society* 10 (6): 789–817.
Easterling, K. (2004) The new orgman: logistics as an organizing principle of contemporary cities. In S. Graham (ed.) *The Cybercities Reader*. London: Routledge, pp. 179–184.
Elden, S. (2007) Governmentality, calculation, territory. *Environment and Planning D: Society and Space* (25): 562–580.
Foucault, M. (1986) Of other spaces. *Diacritics* 16 (1): 22–27.
Foucault, M. (2009) *Security, Territory, Population: Lectures at the Collège de France 1977–1978*, trans. Graham Burchell. New York: Picador.
Gordin, M., Tilley, H. and Prakash, G. (2010) Introduction: utopia and dystopia beyond space and time. In M. Gordin, H. Tilley and G. Prakash (eds) *Utopia/Dystopia: Conditions of Historical Possibility*. Princeton, NJ: Princeton University Press, pp. 1–20.
Hacking, I. (1990) *The Taming of Chance*. Cambridge, MA: Cambridge University Press.
Halpern, O. (2005) Dreams for our perceptual present: temporality, storage, and interactivity in cybernetics. *Configurations* 13 (2): 283–319.
Haraway, D. (1991) A cyborg manifesto: science, technology, and socialist-feminism in the late twentieth century. In D. Haraway (ed.) *Simians, Cyborgs and Women: The Reinvention of Nature*. New York: Routledge, pp. 149–181.
Ho Lee, S., Hoon Han, J., Taik Leem, Y. and Yigitcanlar, T. (2008) Towards ubiquitous city: concept, planning, and experiences in the Republic of Korea. In T. Yigitcanlar, K. Velibeyoglu and S. Baum (eds) *Knowledge-Based Urban Development: Planning and Applications in the Information Era*. Hershey, PA: IGI Global, Information Science Reference, pp. 148–169.
Hollands, R. G. (2008) Will the real smart city please stand up? Intelligent, progressive or entrepreneurial? *City* 12 (3): 303–320.
Kitchin, R. and Dodge, M. (2011) *Code/Space. Software and Everyday Life*. Cambridge, MA: MIT Press.
Knight, F. (1921) *Risk, Uncertainty, and Profit* [online]. Boston, MA: Hart, Schaffner & Marx; New York: Houghton Mifflin Co. Available at: www.econlib.org/library/Knight/knRUP1.html#Pt.I,Ch.I Section I.I.26 [Accessed 24 May 2015].
Koolhaas, R. and Mau, B. (1995) *S, M, L, XL*. New York: The Monacelli Press.
Lakoff, A. (2006) Preparing for the next emergency. Working Paper, Laboratory for the Anthropology of the Contemporary. Berkeley, CA.
LeCavalier, J. (2010) All those numbers: logistics, territory and Walmart [online]. *Places: Forum of Design for the Public Realm* (May). Available at: http://places.designobserver.com/feature/walmart-logistics/13598/ [Accessed 25 May 2015].

Lindsay, G. (2010) The new new urbanism. *Fast Company* (February): 88–95.
Lindsay, G. and Kasarda, J. (2011) *Aerotropolis: The Way We'll Live Next*. New York: Farrar, Straus and Giroux.
Lohr, S. (2012) Big data's impact on the world [online]. *New York Times*, 11 February. Available at: www.nytimes.com/2012/02/12/sunday-review/big-datas-impact-in-the-world.html?pagewanted=all [Accessed 25 May 2015].
MacKenzie, D. (2006) *An Engine, Not a Camera: How Financial Models Shape Markets*. Cambridge, MA: MIT Press.
Mbembe, A. (2004) Aesthetics of superfluity. *Public Culture* 16 (3): 373–405.
Mitchell, W. (1996) *City of Bits*. Cambridge, MA: MIT Press.
Rosenau, H. (1983) *The Ideal City: Its Architectural Evolution in Europe*. New York: Methuen & Co.
Spade, D. (2011) *Normal Life*. New York: South End Press.
Terranova, T. (2004) *Network Culture*. London: Pluto Press.
Weiser, M. (1991) The computer for the 21st century. *Scientific American* (September): 94–104.

10

BEYOND THE CORPORATE SMART CITY?

Glimpses of other possibilities of smartness

Robert G. Hollands

Introduction

Urban development led by the application of information communication technologies (ICTs) has emerged as an important discourse in relation to the future growth, efficiency and prosperity of cities. Entire cities based on smart principles, are currently being constructed in Asia and the Arab world by giant corporate IT, engineering and building firms, while smart initiatives have become commonplace across the USA, Europe and Scandinavia in the last decade. Allegedly motivated by population flows, cities as economic growth hubs and environmental concerns, the smart city is currently being constructed as the solution to many urban problems, including crime, traffic congestion, inefficient services and economic stagnation, promising prosperity and healthy lifestyles for all.

It is counter-intuitive to argue against the idea of a smart city (though for recent critiques see Vanolo, 2014; de Lange and de Waal, 2013; Townsend, 2013, Hemment and Townsend, 2013; Greenfield, 2012; and for an early critique see Hollands, 2008). And there is little doubt that ICTs are significantly transforming urban life (though this is hardly a new idea; see Graham and Marvin, 1995). Despite the idea's inherent positivity, in a recent commentary the renowned urban sociologist Richard Sennett has questioned the logic of the smart city and the largely accepted notion that we should increasingly rely on the use of digital technology to plan our urban environment. Using examples like Masdar, United Arab Emirates (UAE), and Songdo, South Korea, Sennett (2012: online) suggests that the 'danger now is that this information-rich city may do nothing to help people think for themselves or communicate well with one another'.

This critical remark reminds us that there is a series of underlying questions about the 'self-congratulatory' nature of the smart city (see Hollands, 2008), and how ideas about this new urban panacea are currently being promulgated, and by whom. What do we actually mean by the term, and precisely what elements go into making up a smart city? What underlying 'ideological' assumptions are made by invoking the concept, and what are its key contradictions? Who, and what, is driving our preoccupation with the smart city, and who stands to gain and lose in the race towards such an urban future? And finally, are there other, more cooperative and participatory uses of new technology that show glimpses of another kind of smartness that might provide a counterpoint to current conceptions?

The main argument of this chapter is twofold. First, that the idea of the smart city continues to be a highly ideological concept (Hollands, 2008), hiding certain issues and problems from view, while assuming that information technology can automatically make cities more economically prosperous and equal, more efficiently governed and less environmentally wasteful. Second, the way in which this urban panacea is increasingly being packaged and promoted is that it can only be effectively delivered through a corporate vision of smartness (see Söderström et al., 2014), in conjunction with an entrepreneurial form of urban governance (Harvey, 1989), and a largely compliant and accommodating citizenry (Gabrys, 2014; see Chapter 6). This argument entails making a twofold intervention into the debate surrounding the rise of this corporate-orientated smart city. First, the chapter looks critically at how we currently understand the smart city, specifically focusing on the rising trend towards corporate and entrepreneurial governance versions. A second form of intervention concerns considering smartness from a different perspective, emanating from small-scale and fledgling examples of participatory and people-power types of smart initiative (Brickstarter, n.d.; Chatterton, 2013; de Lange and de Waal, 2012; Radywyla and Biggs, 2013). The reason for discussing these few examples is not to suggest that they pose a readymade alternative to the corporate vision. Rather, they offer only a glimpse into different and more human versions of smartness, using technology to realise progressive ideas, rather than see the technology as progressive in and of itself (de Lange and de Waal, 2013).

Understanding the smart city concept: visions, elements, and trends

Examples of transformed lives in smart cities come from a variety of sources, namely, IT corporate websites, futuristic films and academic and policy-making circles. Fujitsu, a leading Japanese ICT company says it is 'striving to leverage ICT to create a society where people's lives are prosperous and more secure' (Fujitsu, n.d.: online), while Cisco, which has been involved as the IT partner in the creation of the first smart city from scratch in South Korea, Songdo, says on its website that the city will 'create the ultimate lifestyle and work experience' (Cisco, n.d.: online). The ICT powerhouse IBM on its website claims that: 'Smart growth can lead to safe neighbours, quality schools, affordable housing and traffic that flows. It's all possible' (IBM, n.d.: online). Popular cultural images in the form of futuristic films are less flattering and more concerned about the negative impact technology can have on our urban lives. While the *Terminator* series of movies is perhaps the most obvious dystopic representation of what happens when the machines (computers) take over, films like *Equilibrium*, *Bladerunner* and *Minority Report* also raise important issues about information technology and its relationship to urban privacy, security and hyper-consumerism.

Discussions about smart cities in academic circles are of course more varied, diverse and complex than these corporate utopian visions or cinematic false dawns. Part of this more complex understanding comes down to the varied ways the term has been employed or linked to related concepts. For example, while the adjective 'smart' clearly implies some kind of positive urban-based technological innovation and change via ICT, analogous to the 'wired', 'digital', 'informational' or 'intelligent' city, it is not, as has been argued elsewhere, exactly synonymous with these terms (Hollands, 2008). More recently, some writers have begun to talk about the ubiquitous or 'u-city', where smart technology is completely embedded in the urban fabric and all urban systems become linked through IT advancements (Anttiroiko, 2013). Smart initiatives have also been discussed in relation to a range of ideas, including the learning or knowledge city (McFarlane, 2011; Campbell, 2012), their link to creative cities (Florida, 2010) and, more recently, open data sharing in cities (Bates, 2013), urban transitions to low carbon output (Bulkeley et al., 2010) and increased discussion about green cities as smart (Joss et al., 2013).

This diversity of ideas creates certain conceptual problems in discussing smart cities, as different writers invoke quite varied aspects in their definition of the term. For example, some view smartness almost exclusively as technology and hardware (see Moyser, 2013), while others emphasise good urban governance and services (Comstock, 2012). Still others use definitions that give primacy to smart technologies that reduce our energy consumption and environmental footprint (Cohen, 2012). The Centre of Regional Science (2007) utilises a range of measures in ranking smart cities, including six main smart characteristics – economy, people, governance, mobility, environment and living – possessing 31 factors and having 74 indicators that they can be measured by.

Effectively a smart city is made up of information technology devices, industry and business, governance and urban services, neighbourhoods, housing and people, education, buildings, lifestyle, transport and the environment. Because it is made up of such a diverse range of things, the smart city idea can inadvertently bring together different aspects of urban life that do not necessarily belong together, hiding some things and bringing others to the fore (see Hollands, 2008). This is essentially what I mean by arguing that the concept is an ideological one (following Marx's basic idea that ideology clouds or hides its real or true meaning). For example, the unspoken assumption in the corporate quotations above is that the application of information technology in cities will automatically benefit everyone, with prosperity and wealth being shared by all. Or that we all roughly share the same kind of smart city vision. Others have argued that smartness is ideological in that sense the it can become a self-imposed label, a marketing device for city branding (Hollands, 2008).

What new trends are evident in the literature on smart cities today? First, it is clear that there is still a plethora of examples of smart cities and smart city initiatives that one could highlight, which implies that it continues to be a significant urban development. Popular examples abound on the Internet, including large-scale grand plans like Singapore's iN2015 (intelligent nation) project, Songdo, South Korea's purpose-built, globally competitive, high-tech, environmentally sustainable business city, and Guangzhou Knowledge City in China, designed to attract talent, skilled workers and knowledge-based industries. In Scandinavia and elsewhere in Europe, Helsinki (Finland) and 'Intelligent' Thessaloniki (Greece) are held up as examples of encouraging the development of new mobile applications utilising open data and using IT to increase competitiveness and sustainability, respectively (Komninos *et al.*, 2013). In Europe, Barcelona continues to be renowned for its Smart City Model, while the Amsterdam Smart City initiative is held up as the example of how to retrofit a city to improve living and economic conditions and reduce carbon emissions (Kirby, 2013). In the UK, Manchester's FutureEverything programme is meant to make it the world's first 'open data' city, while Glasgow has recently won £24m from the government to demonstrate how a smart city of the future might operate (Wakefield, 2013). Even struggling cities like Sunderland are getting in on the act, with the CEO of the council saying: 'I see this opportunity through smarter cities as being the next revolution' (Kirby, 2013: online).

The question is what does this proliferation of examples actually tell us? For although we might be increasingly surrounded by the discourse of smartness, the development of initiatives is perhaps more uneven and slower than once envisaged (see Schelmetic, 2011). It is also the case that there are critical differences in the *scales* and *types* of smart projects adopted. For example, Songdo is a ten-year, $40b urban development the size of Boston, while Stratford, Ontario (population 32,000) has been named one of the world's Top 7 Intelligent Communities by the Intelligent Community Forum three years in a row (see StratfordSmartCity, 2013). Many of the smart city examples existing on the Internet are specific and varied initiatives rather than full-blown programmes, and there are very different national and international patterns of smart development, all of which need further unpacking.

Many accounts of smart cities also cite the urgent need for environmental solutions, as urban areas consume 75 per cent of the world's energy and are responsible for 80 per cent of greenhouse gas emissions. Pike Research, a tracking organisation, suggests that more than 50 per cent of projects they are assessing have focused on innovations in transportation and urban mobility (Navigant Research, n.d.), and ABI Research estimated that smart grids accounted for 36 per cent of total smart city expenditures in 2011 (Korzeniowski, 2012). The European Smart Cities Initiative is also focused on the sustainability issues of cities and, more specifically, on their energy systems (European Commission, 2010), as are many Scandinavian projects, with Copenhagen aiming to be the world's first carbon neutral capital (Copenhagen Cleantech Cluster, n.d.). Many of the mega-developments in Asia and the UAE are based on environmental sustainability as their rationale, though one needs to factor in construction energy costs to build them.

Despite this latter progressive-sounding credential, and the claim on the Cisco website that the 'Songdo project is a model for smart cities around the globe' (Cisco, n.d.: online), perhaps as illuminating is the comment from Stan Gale, Chairman, Gale International, that: 'The concept behind it is that this would become the central focal point and a main alternative for large-scale companies looking to do business in Japan, China and Korea' (Cisco, n.d.: online). Essentially, Songdo is a giant business park, not a city *per se*. The development is set out in effect to produce an ideal corporate 'lifestyle and business experience' (Cisco, n.d.: online), with the idea that people can come in from overseas and live, work and enjoy leisure completely within the confines of pure corporate spaces. Everyday urban life comes complete with home/office/educational/government interface systems (unfortunately called Telepresence), a Jack Nicklaus-designed golf course and corporate shopping areas.

This is hardly a one-off experiment. Once New Songdo City is finished, its builder plans to roll out 20 new cities across China and India, presumably with Cisco in tow to build the city's central brains. Other giant corporations also see the smart city idea as both a driver of urban change and a source of future profits. For example, Fujitsu, a leading Japanese ICT company with revenues of US$54b, argues on the company website:

> The Fujitsu Group will promote smart cities as an impetus for social change. In line with its long-term vision of realizing a Human Centric Intelligent Society, the Fujitsu Group is striving to leverage ICT to create a society where people's lives are prosperous and more secure.
>
> *(Fujitsu, n.d.: online)*

In 2008, in the midst of the banking crisis, the high-tech giant IBM rebranded itself with a Smarter Planet initiative as a lynch-pin of its growth strategy, holding 100 Smarter City Forums around the world, and now claims to be involved in around 2,000 smart projects worldwide. This strategy has clearly paid off, generating $3b (double-digit growth in this area), from nearly 6,000 clients. Currently 25 per cent of IBM's operations are in smart area, and this is set to double over the next few years (IBM, n.d.).

Numerous other large-scale smart city projects exist, namely Masdar, in the UAE, and PlanIT Valley, in Porto, Portugal. While the UK has nothing on this scale, more discrete examples of it are beginning to emerge. LandProp, a property offshoot of InterIkea, the parent company of the well-known furniture store, is currently developing a mini-city called Strand East in East London (Beanland, 2012). Urban writer Anna Minton, in her fascinating book *Ground Control* (2009), has been arguing that public spaces in many UK cities have been increasing privatised and turned over to corporate control, with ill effects. Other critics are unhappy about the idea of future smart cities growing up entirely around corporate power and money, and stress that it is social and urban development that happens after the technology is put in that is crucial to the liveability and sustainability of these cities (Schelmetic, 2011).

No less significant examples of corporate influence on urban development connected to the use of smart technology are in the area of advertising and consumerism. Akin to the futuristic movie, *Minority Report*, where Tom Cruise runs through a mall as the advertisements around him change to tailor exactly to his tastes, Immersive Labs, a start-up tech company, will shortly trial its first camera-enhanced 'smart signs', equipping billboards and retail signage in places like airports, malls and retail stores with the ability to compute what type of consumer is looking back: male or female, young or old, a sports fan or a pet owner (Curry, 2011). Engineers in Japan from the electronics company NEC have already developed a billboard that is capable of identifying a shopper's age and gender through facial recognition software, as they walk past, to offer them products that are more accurately suited to them (Gray, 2010).

Why are we seeing a trend whereby our cities are increasing becoming a backdrop to corporate advertising and the privatisation of public space? And why are city leaders eager to hand over cash and control to business-led smart urban development? The urban geographer David Harvey (1989) noted a significant global shift in forms of city governance back in the mid-1980s, away from a managerial welfare orientation to one of urban entrepreneurialism. Strapped for cash, cities began to compete with one another in attracting in global capital and marketing themselves as world-leading cultural, creative or smart brand cities. With the global banking crisis of 2008, followed by a nearly worldwide politics of austerity, this governance trend has continued with an increased emphasis on efficiency savings, privatisation and the promise of a high-tech future. As corporate ICT companies themselves have noted:

> [I]n the 21st century, cities compete globally to attract both citizens and businesses. A city's attractiveness is directly related to its ability to offer the basic services that support growth opportunities, build economic value and create competitive differentiation .... They are looking for smarter cities.
>
> *(IBM, 2012: online)*

There are different international patterns of entrepreneurial governance, privatisation and corporatisation here, all of which impact on the scale and direction smart initiatives take. For instance, while North American and European governance models are still entrepreneurial in the Harvey (1989) sense, democratic controls and privacy/security concerns may mean that there are more cautious and nuanced examples of cooperation between city governments, citizens and business. Hudson (2010) also talks about the notion of 'resilient regions' and discusses some of the ways in which places can begin to push against the negative effects of new-liberal capitalist development. However, as Anttiroiko (2013: 7–8) has argued, in places like Japan and South Korea, there is much more direct collusion of corporate and local government interests, a longer history of the privatisation of national telecommunications systems and more examples of all-encompassing ubiquitous smart developments. For example, Korea Telecom, involved in Songdo, was once a public corporation but was privatised in 2000 and then became a major driver of the u-city concept which emerged in political circles there in 2004 (Anttiroiko, 2013: 8). A similar form of privatisation occurred prior to the Singaporean government launching the Intelligent Nation 2015 (iN2015) programme (see Hollands, 2008: 312), whose aim is to transform the country into an intelligent nation and a global city.

A key question raised here about information technology and public–private smart partnerships is: who gains and who loses through such arrangements? Regarding the creation of Smart Grids, for example, putting the necessary IT infrastructure in place requires a significant investment. The Stockholm Royal Seaport project, for example, came with a preliminary price tag of $2.9m, with the Swedish Energy Agency paying $1.2m and Vinnova, the Swedish Governmental Agency for Innovation Systems, contributing another $700,000, while the remaining funding (around a third) came from the participating vendors (Korzeniowski, 2012). Another example concerns Sunderland Council's £5.7m investment in the Sunderland Computing Cloud. While they suggested that they would recoup their investment in five years' time thanks to 'efficiencies in services' and through a boost to the IT economy in the region (Parnell, 2011), at the same time the council had cut 1,500 jobs since 2009 and in 2013 was making cuts of £37m (including £3.8m to child services and £5.1m to health, housing and adult services; see Sunderland Echo, 2013). In the wake of urban austerity, it is unclear to what extent local and national governments can continue to foot the bill for public–private partnerships and effectively subsidise private industry in the smart field, when councils cannot even provide basic urban services for the majority of people who live in cities (Korzeniowski, 2012; Hoornweg, 2011).

A final question, not really raised in the literature, is to what extent the corporate entrepreneurial smart city 'is in its fragmented mode a new way of building functionally sophisticated enclaves into society, which tends to serve mainly high value adding activities and high income people?' (Anttiroiko, 2013: 13). Serious urban problems like poverty, inequality and discrimination appear to be largely absent from these neoliberal urban visions and projects, and there appears to be little or no recognition that smart developments might contribute negatively to social polarisation in cities – what Graham and Marvin (2001) have referred to as splintering urbanism (see also Graham, 2002). In the main, most smart initiatives envisioned here come from either corporations or urban governments, not from actual people who live and work in cities. This lack of concern with democratic decision-making and real citizen involvement, participation and control of most smart city projects has led some urban critics to search for different ways to think about smartness and to explore smaller-scale, community-based and more socially progressive uses of new technologies.

Interventions in the corporate smart city: glimpses of possibilities?

While neither definitions of or practices surrounding smart cities and smart initiatives are a monolith, the argument made so far is that there is a growing tendency for them to be technologically led, corporately influenced and tied to competitive models of the entrepreneurial city. This is especially the case with regard to Asian models of the corporate ubiquitous city, although as Provoost (2012) has argued, smaller-scale models of this type are also being trialled in Europe. Previous research into a number of smart city initiatives in Europe and North America showed that a significant proportion were undertaken by city governments for urban marketing/branding purposes (Hollands, 2008), rather than being citizen-led. This is not to suggest there are no well-meaning and progressive initiatives out there, designed to solve pressing urban problems related to things like urban decline, transport issues or making cities more carbon neutral. However, there exist no large-scale alternative smart city models, partly because most cities have generally embraced a pro-business and entrepreneurial governance model of urban development, and hence are subject to many of the same kinds of criticism that might be made of the more extreme, corporately organised u-city type (Anttiroiko, 2013).

Another problem in defining what might be meant by 'alternative' is whether we are talking about future visions or immediate practicalities. Generalised alternative urban visions, for example, tend to be rather vague and utopian models arguing for sustainable resources, not a money-based, world capitalist economy (see the ideas of the Zeitgeist Movement and the Venus Project, for example). Similarly, there exist tactical technologically based movements, such as the International Pirate Party, who have campaigned for open copyright and the use of social media to get issue petitions and consensus-based decision-making on the table. Still others have emphasised challenging the corporate grip on information technology through the provision of free software (Kelty, 2008), or politically challenging the status quo by creating loosely associated networks of 'hacktivists' and 'cyber guerrillas' (like the group Anonymous, amongst others; see Ronson, 2013). While the difficulty facing groups like Zeitgeist and Venus is the lack of feasibility of a resource-based approach in light of the dominance of neoliberal global capitalism, the weakness of the second approach is ironically its exclusive use of technology as a basis for political action.

Perhaps more instructive would be to examine a range of more modest and small-scale socio-technological interventions that contrast with the corporate smart city, and which might begin to help us envisage a different way of thinking about and 'doing' smartness. Before turning to a brief discussion of four examples, it might be useful to outline a few basic differentiating principles. For instance, one of the most important principles to start with here is the need to begin to move away from the idea that technological solutions, in and of themselves, are the only viable (and easiest) way to solve our many urban problems (Hoornweg, 2011; Hill, 2013). Second, we need to shift the debate about smart cities towards the *raison d'être* of cities – the people and citizens who live in them (Hill, 2013). Third, as de Lange and de Waal (2013: online) have argued, one of the key elements of imagining a different kind of smartness concerns ideas about ownership, not limited to proprietorship, but rather, in their words, 'how to engage and empower citizens to act on complex collective urban problems'. This involves starting with urban citizens taking responsibility and acting collectively, but also raises issues of social learning and social cooperation. For Hudson (2010), this also requires using human capabilities to reduce social risks, while at the same time affording socially useful and environmentally enhancing activity much greater recognition and significance.

There are, of course, many examples that might fulfil most aspects mentioned here, and the difficulty is always which projects to highlight. The brief discussion of four cases below is not meant to be in any way exhaustive or comprehensive, but rather instructive. Similarly, it is important to understand these examples in the context of the principles just mentioned, rather than writing them off as 'sustainability projects'. They all use technology in some way to help solve urban problems – however, its use supplements and supports progressive and smart solutions based on collective ideas, action and resilience, rather than starting with the technology as the driving force (de Lange and de Waal, 2013).

Many of these ideals are contained in the fledgling urban crowd-source idea called Brickstarter. According to their website, they are 'sketching a system that would enable everyday people, using everyday technology and culture, to articulate and progress sustainable ideas about their community' (Brickstarter, n.d.: online). The general philosophy behind this new initiative is to utilise social media to be more responsive, representative and educative in transforming grassroots urban proposals into viable projects (what they call YIMBY, or yes in my back yard). One commentator has suggested that it could make 'citizen-based urban planning a reality' (McGuirk, 2012: online). Their prototype IT platform invites and advises groups on how to negotiate their way through what Brickstarter calls the 'dark matter' of local city planning, and, more important, how they might be able to fund such a project, through a kind of urban crowd-funding/sourcing platform. While there remain issues over the eventual operationalisation of the Brickstarter platform (only a basic prototype exists; see Boyer and Hill, 2013), not to mention the problems of involving poorer urban dwellers and of crowd-sourcing becoming part of the neoliberal cost-cutting agenda (McGuirk, 2012), there are also distinct possibilities raised here regarding citizen involvement in urban issues.

An actually existing project combining an energy-efficient technology with a community focus is the Leeds housing project LILAC (Low Impact Living Affordable Community; Figure 10.1). In an effort to solve the twin problems of affordable yet ecologically sustainable housing, as well as encourage cooperative community-based living, LILAC has become the UK's first Mutual Home Ownership Scheme. Funded by an eco-friendly bank and a grant from the Homes and Communities Agency (on a site sold to them at a reduced rate by the council), resident households pay 35 per cent of their income into a trust, thereby acquiring equity shares, enabling even those with incomes of £15,000 to get on the housing ladder (Wainwright, 2013). In terms of using sustainable technology, the project aims to be as low carbon as possible as the houses are of wooden construction with straw-bale insulation, have rainwater collection, energy-efficient heating, minimal car spaces and a shared tool shed. Community-wise, LILAC has been designed with communal values in mind, with a common area with shared kitchen, laundry, workshop, meeting/function room, in addition to which each unit has its own allotment

FIGURE 10.1 An ecological community: LILAC co-housing project opening party, September 2013
Source: LILAC (used by permission)

to grow food (see Chatterton, 2013). While the project no doubt waded through a lot of local authority 'red tape' to get off the ground, it is an excellent example of a grassroots initiative, where people, not corporations or politicians, control their urban lives, and is a potential model for providing affordable and sustainable housing in other areas of the UK. Recently they have won two city architectural awards.

De Lange and de Waal (2013), on the other hand, do not focus much on community initiatives or forums but what they call 'networked publics', and they examine a range of examples here from data commons and media art projects to DIY urban design. Regarding this last category here, they argue that digital media can help enable collective action. The example they discuss is an interesting project called *Face your World*, set up by two artists, which invited young people and neighbours living in an Amsterdam neighbourhood to collaborate in producing a virtual vision of their local park, which they persuaded the local government to adopt in place of their own plan (de Lange and De Waal, 2013). In their longer e-publication on ownership in the hybrid city, de Lange and de Waal (2012: 25) suggest that the 'project brought together a variety of urban issues including urban regeneration, practical education, community participation and art in public space'.

A final example combining information technology and social media with sustainability is 596 Acres, a project designed to bring Brooklyn's 596 acres of public-owned land into common use by a range of community groups and individuals. Its online IT platform, effectively a 'knowledge commons', has been crucial in building this intervention, connecting people to each other, matching skills and sharing experience and information about how to transform vacant lots into sustainable growing plots (Radywyl and Biggs, 2013). The implications of projects like this are not, however, just about using technology for progressive politics or developing skills, but are also crucial for building social capital, community and urban sustainability. Eizenberg's (2013) excellent book *From the Ground Up* is a study of 650 community gardens in New York City, which are managed collectively by some of the city's least well-off residents for purposes of horticulture, recreation, social gatherings and artistic and cultural events. She argues that these community gardens create not only ecological spaces but 'organic urban residents' and actors, making a city in their own image. What is being argued here is that alternative smart projects are smart by virtue of solving a number of urban problems simultaneously (community spirit, social capital, sustainability, availability of fresh and affordable food, etc.), rather than just being technological planning devices.

All of these examples exemplify not just a 'right to use technology', which is precisely where many smart city initiatives stop, but rather the right to shape the city using human initiative *and* technology for social purposes to make our cities better and more sustainable. This idea has a number of implications. First, smart initiatives do not have to be large-scale and costly or always motivated by corporate profit-seeking or competitive city brand-makers. Selling high-tech ideas and hardware to cities is expensive and may only benefit the few, argues Hoornweg (2011). Second, as Michael Andrew McAdams (2013: online) suggests: 'It would seem obvious, but a "smart city" must be inhabited by "smart people"' (see also Hemment and Townsend, 2013). This requires, in his view, open access to an excellent system of education, including at university level, in order for people to engage more democratically with intelligent technology. Similarly while there have been suggestive discussions about smartcitizens (Hemment and Townsend, 2013) – as well as ideas about the city as a learning machine (McFarlane, 2011) and urban knowledge hubs (Campbell, 2012) – in the main, existing corporate smart city models tend to see citizens as a barrier to the implementation of smart technology (because of technological ignorance or lack of education), or just as another resource, as in human-capital-type approaches. For the citizen, smartness becomes reduced to a form of smart mentality (Vanolo, 2014), simply adopting the right frame of mind to accept and cope with the inevitability of urban technological change. Hoornweg (2011: online) argues: 'At its core a smart city is a welcoming, inclusive city, an open city .... A smart city listens – and tries to give voice to everyone.' We need to ask if current visions of corporately led smart cities actually do this, and if not, consider what other interventions need to be adopted if they are to begin to move in this direction.

Conclusion

As Sennett (2012: online) states: 'We want cities that work well enough, but are open to the shifts, uncertainties, and mess which are real life.' Contrary to dominant representations that urban development through the application of ICTs is both a positive and inevitable trend, the smart city concept raises more questions than it answers. The suggestion by giant IT consortiums that we need to become technologically smarter now to save our cities, and consider the social consequences later, is highly pre-emptive, not to mention ideological. We should be wary of corporately inspired smart scenarios where urban problems have all been solved by technology and all of the inhabitants are happy and prosperous, however tantalising this vision is. Underlying this idea is a more manipulative notion that cities are just 'machines for making money out of', or that global competitiveness between cities will automatically make them better places to live.

For too long, smart city discourses have been ignorant as to how cities actually work sociologically and politically, and the fact that they are made up of a complex and diverse set of dynamics and conflicts (Harvey, 2012). They also fail to ask important questions about urban life: why are most cities unequal places; what economic system created the current ecological conditions; how can cities organically develop and real communities form; and, what is the good or fair city? (Toderian, 2012). We need better socio-political understandings of the city, and more novel approaches emphasising the need to see urban technological transformation within a wider social, political, economic, cultural and organisational context. And we need to engage very much with real-time, citizen-led smart initiatives and cases studies, looking critically and carefully at the policy process, driving forces, power and sociological context.

Many of our major urban problems are not technological, but social, like poverty and inequality, and have been exacerbated, not solved, by corporate privatisation and city branding strategies (Harvey, 2012). Additionally, there has been little room for people power, democratic debate and citizen rights in many discussions of the smart city. Their role has too often been limited to being in the right frame of mind to accept the inevitability of the smart city – i.e. to develop a smartmentality to cope with urban technological change. As Anttiroiko asks:

> Here, the critical question is whether u-city really benefits us all, or is it ultimately a capital affirmative endeavour of which construction companies and UbiTech firms reap the most benefit, public sector carries major risks through their support schemes and public investments, and people are made to adjust to a new technologically mediated mode of urban life, without much room for choices of their own.
>
> *(Anttiroiko, 2013: 13)*

Urban life, as urban sociology over the past century has shown us, is a multifaceted and complex thing. Problems like urban poverty, discrimination, inequality and social polarisation, issues like neighbourhood and community decline, crime and neglect, and even environmental problems like traffic congestion and recycling, have important social, political and cultural dimensions, and will not be ameliorated solely by simple technological solutions or more sophisticated data gathering. This is the paradox faced by any smart initiative – corporate or otherwise. Participation-based and citizen-run interventions into the smart city give us no more than glimpses of what is and might be possible if IT was used progressively and in the service of urban dwellers, rather than for simply efficient high-tech 'quick fixes' and corporate profit-making activities. The question is, can we afford not to consider different ideas of smartness beyond the corporate form?

References

Anttiroiko, A.-V. (2013) U-cities reshaping our future: reflections on ubiquitous infrastructure as an enabler of smart urban development. *AI & Society* 28: 491–507.

Bates, J. (2013) The domestication of open government data advocacy in the United Kingdom: a neo-Gramscian analysis. *Policy & Internet* 5 (1): 118–137.

Beanland, J. (2012) London's newest development: the rise of the Ikea city [online]. *Independent*, 4 October. Available at: www.independent.co.uk/news/uk/home-news/londons-newest-development-the-rise-of-the-ikea-city-8196429.html [Accessed 15 May 2013].

Boyer, B. and Hill, D. (2013) Brickstarter [online]. Available at: www.helsinkidesignlab.org/peoplepods/themes/hdl/downloads/Brickstarter.pdf [Accessed May 20, 2015].

Brickstarter (n.d.) An introduction [online]. Available at: http://brickstarter.org/an-introduction-to-brickstarter/ [Accessed May 14, 2013].

Bulkeley, H. A., Castan Broto, V., Hodson, M. and Marvin, S. (eds) (2010) *Cities and Low Carbon Transitions*. London: Routledge.

Campbell, T. (2012) *Beyond Smart Cities: How Cities Network, Learn and Innovate*. Abingdon: Earthscan.

Centre of Regional Science (2007) *Smart Cities: Ranking of European Medium-Sized Cities* [online]. Available at: www.smart-cities.eu/download/smart_cities_final_report.pdf [Accessed 19 May 2013].

Chatterton, P. (2013) Towards an agenda for post-carbon cities: lessons from LILAC, the UK's first ecological, affordable, cohousing community. *International Journal of Urban and Regional Research* 37 (5): 1654–1674.

Cisco (n.d.) *Cities of the Future: Songdo, South Korea* [online]. Available at: http://newsroom.cisco.com/songdo [Accessed 14 May 2013].

Cohen, B. (2012) The top 10 smart cities on the planet [online]. *CoExist*, 17 January. Available at: www.fastcoexist.com/1679127/the-top-10-smart-cities-on-the-planet [Accessed 17 April 2013].

Comstock, M. (2012) What is a smart city and how can a city boost its IQ? [online]. blogs.worldbank.org. Available at: http://blogs.worldbank.org/sustainablecities/node/576?cid=EXT_TWBN_D_EXT [Accessed 14 May 2013].

Copenhagen Cleantech Cluster (n.d.) Copenhagen as a carbon-neutral smart city [online]. Available at: www.cphcleantech.com/ccj2-copenhagenasacarbonneutralsmartcity [Accessed 14 May 2013].

Curry, C. (2011) Inspired by 'Minority Report', billboards recognize faces [online]. *ABC News*, 20 September. Available at: http://abcnews.go.com/blogs/technology/2011/09/inspired-by-minority-report-billboards-recognize-faces/ [Accessed 15 May 2013].

de Lange, M. and de Waal, M. (2012) *Ownership in the Hybrid City*. Rotterdam: Virtueel Platform.

de Lange, M. and de Waal, M. (2013) Owning the city: new media and citizen engagement in urban design [online]. *First Monday* 18 (11), November. Available at http://firstmonday.org/ojs/index.php/fm/article/view/4954/3786 [Accessed February 7, 2014].

Eizenberg, E. (2013) *From the Ground Up: Community Gardens in New York City and the Politics of Spatial Transformation*. Farnham: Ashgate.

Eurocities (2012) *Statement on Smart Cities and Communities Communication* [online]. Available at: http://nws.eurocities.eu/MediaShell/media/EUROCITIES%20Statement%20on%20Smart%20Cities%20Communication%20Oct%202012.pdf [Accessed May 9, 2013].

European Commission (2010) Strategic energy technology plan information system: European Initiative on Smart Cities [online]. Available at: http://setis.ec.europa.eu/initiatives/technologyroadmap/european-initiativeon-smart-cities [Accessed 13 May 2013].

Florida, R. (2010) Smart work and smart cities pay [online]. *The Atlantic*, 23 April. Available at: www.theatlantic.com/national/archive/2010/04/smart-work-and-smart-cities-pay/39393/ [Accessed 9 May, 2013].

Fujitsu (n.d.) Fujitsu envisages smart cities [online]. Available at: www.fujitsu.com/global/about/responsibility/feature/2012/smartcity/ [Accessed 14 May 2013].

Gabrys, J. (2014) Programming environments: environmentality and citizen sensing in the smart city. *Environment and Planning D: Society and Space* 32(1): 30–48.

Graham, S. (2002) Bridging urban digital divides? Urban polarisation and information and communications technologies (ICTs). *Urban Studies* 39 (1): 33–56.

Graham, S. and Marvin, S. (1995) *Telecommunications and the City: Electronic Spaces, Urban Places*. London: Routledge.

Graham, S. and Marvin, S. (2001) *Splintering Urbanism: Networked Infrastructures, Technological Mobilities and the Urban Condition*. London: Routledge.

Greenfield, A. (2012) Week 61: spontaneous order (and value) from the bottom up [online]. Available at: http://urbanscale.org/news/ [Accessed 14 May 2013].

Gray, A. (2010) *Minority Report*-style advertising billboards to target consumers [online]. *Telegraph*, 1 August. Available at: www.telegraph.co.uk/technology/news/7920057/Minority-Report-style-advertising-billboards-to-target-consumers.html [Accessed 15 May 2013].

Haque, U. (2012) Surely there's a smarter approach to smart cities? [online]. *Wired*, 17 April. Available at: www.wired.co.uk/news/archive/2012-04/17/potential-of-smarter-cities-beyond-ibm-and-cisco [Accessed 14 May 2013].

Harvey, D. (1989) From managerialism to entrepreneurialism: the transformation in urban governance in late capitalism. *Geografiska Annale* 71B (1): 3–17.

Harvey, D. (2012) *Rebel Cities: From the Right to the City to the Urban Revolution*. London: Verso.

Hemment, D. and Townsend, A. (2013) *Smart Citizens*. Manchester: FutureEverything. Available at: http://futureeverything.org/wp-content/uploads/2014/03/smartcitizens.pdf [Accessed May 9. 2014].

Hill, D. (2013) On the smart city; or, a 'manifesto' for smart citizens instead [online]. *City of Sound*, February. Available at: www.cityofsound.com/blog/2013/02/on-the-smart-city-a-call-for-smart-citizens-instead.html [Accessed 14 May 2013].

Hollands, R. (2008) Will the real smart city please stand up? Intelligent, progressive or entrepreneurial? *City* 12 (3): 303–320.

Hoornweg, D. (2011) Smart cities for dummies [online]. blog.worldbank.org. Available at: http://blogs.worldbank.org/sustainablecities/smart-cities-for-dummies [Accessed 15 May 2013].

Hudson, R. (2010) Resilient regions in an uncertain world: wishful thinking or a practical reality? *Cambridge Journal of Regions, Economy and Society* 3: 11–25.

IBM (2012) Smarter more competitive cities [online]. IBM Corporation. Available at: http://public.dhe.ibm.com/common/ssi/ecm/en/pub03003usen/PUB03003USEN.PDF [Accessed 15 May 2013].

IBM (n.d.) Smarter planet [online]. Available at: www-03.ibm.com/ibm/history/ibm100/us/en/icons/smarterplanet/ [Accessed 15 May 2013].

IBM website (n.d.) Smarter cities [online]. Available at: www.ibm.com/smarterplanet/uk/en/smarter_cities/ideas/index.html [Accessed 14 May 2013].

Joss, S., Cowley, R. and Tomozeiu, D. (2013) Towards the 'ubiquitous eco-city': an analysis of the internationalisation of eco-city policy and practice. *Urban Research & Practice* 6 (1): 54–74.

Keeton, R. (2011) *Rising in the East: Contemporary New Towns in Asia*. Amsterdam: Sun Architecture.

Kelty, C. (2008) *Two Bits: The Cultural Significance of Free Software*. London: Duke University Press.

Kirby, T. (2013) City design: transforming tomorrow [online]. *Guardian*, 18 April. Available at: www.guardian.co.uk/smarter-cities/transforming-tomorrow [Accessed 15 May 2013].

Komninos, N., Pallot, M. and Schaffers, H. (eds) (2013) Special issue: Smart cities and the future internet in Europe. *Journal of the Knowledge Economy* 4 (2): 119–134.

Korzeniowski, P. (2012) Smart grid smart cities [online]. *EnergyBiz Magazine*, July/August. Available at: www.energybiz.com/magazine/article/273639/smart-grid-smart-cities [Accessed 14 May 14].

Lima, H.-C. and Jangb, J.-O. (2006) Neo-liberalism in post-crisis South Korea: social conditions and outcomes. *Journal of Contemporary Asia* 36 (4): 442–463.

McAdams, M. A. (2013) A 'smart city' or the 'matrix'? [online]. Available at: www.progressivepress.net/a-smart-city-or-the-matrix/ [Accessed 15 May 2013].

McFarlane, C. (2011) The city as a machine for learning. *Transactions of the Institute of British Geographers* 36: 360–376.

McGuirk, J. (2012) Brickstarter: crowd-funding takes to the streets [online]. *Guardian*, 24 August. Available at: http://webcache.googleusercontent.com/search?q=cache:http://www.theguardian.com/artanddesign/2012/aug/24/brickstarter-crowd-funding [Accessed 10 February 2014].

Minton, A. (2009) *Ground Control: Fear and Happiness in the Twenty-First-Century City*. London: Penguin Books.

Moyser, R. (2013) Defining and benchmarking SMART cities [online]. Available at: www.burohappold.com/blog/article/defining-and-benchmarking-smart-cities-1771/ [Accessed 15 May 2013].

Navigant Research (n.d.) Smart City Tracker 1Q13: global smart city projects by world region, market segment, technology, and application [online]. Available at:]www.navigantresearch.com/research/smart-city-tracker-1q13 [Accessed 15 May 2013].

Parnell, B.-A. (2011) Sunderland hires IBM to build cloud infrastructure [online]. *The Register*, 16 November. Available at: www.theregister.co.uk/2011/11/16/ibm_build_cloud_for_sunderland/ [Accessed 13 February 2014].

Provoost, M. (2012) From welfare city to neoliberal utopia [online]. Strelka Talks. Available at: www.strelka.com/content/michelle-provoost/?lang=en [Accessed 15 May 2013].

Radywyla, N. and Biggs, C. (2013) Reclaiming the commons for urban transformation. *Journal of Cleaner Production* 50 (1): 159–170.

Ronson, J. (2013) Security alert: notes from the frontline of the war in cyberspace [online]. *Guardian*, 4 May. Available at: www.theguardian.com/technology/2013/may/04/security-alert-war-in-cyberspace [Accessed 25 May 2015].

Schelmetic, T. (2011) The rise of the first smart cities [online]. *IMT Green and Clean Journal*, 20 September. Available at: http://news.thomasnet.com/green_clean/2011/09/20/the-rise-of-the-first-smart-cities/ [Accessed 17 April 2013].

Sennett, R. (2012) No one likes a city that's too smart [online]. *Guardian*, 4 December. Available at: www.guardian.co.uk/commentisfree/2012/dec/04/smart-city-rio-songdo-masdar [Accessed 3 January 2013].

Söderström, O., Paasche, T. and Klauser, F. (2014) Smart cities as corporate storytelling. *City* 18: 307–320.

StratfordSmartCity (2013) The Stratford global standard: setting the bar for smart city projects [online]. Available at: http://stratfordsmartcity.ca/2013/03/the-stratford-global-standard-setting-the-bar-for-smart-city-projects/ [Accessed 14 May 2013].

Sunderland Echo (2013) Sunderland Council approves £37million cutbacks [online]. *Sunderland Echo*, 7 March. Available at: www.sunderlandecho.com/news/local/all-news/sunderland-council-approves-37million-cutbacks-1-5472972 [Accessed 13 February 2014].

Toderian, B. (2012) Asking the right questions about Memphis [online]. *Smart City Memphis*, 12 July. Available at: www.smartcitymemphis.com/2012/07/asking-the-right-questions-about-memphis/ [Accessed 15 May 2013].

Townsend, A. (2013) *Smart Cities: Big Data, Civic Hackers, and the Quest for a New Utopia*. New York: W.W. Norton.

Vanolo, A. (2014) Smartmentality: the smart city as disciplinary strategy. *Urban Studies* 51 (5): 883–898.

Wainwright, O. (2013) The future's communal: meet the UK's self-build pioneers [online]. *Guardian*, 7 May. Available at: www.theguardian.com/artanddesign/ 2013/may/07/uk-self-build-pioneers [Accessed 30 June 2014].

Wakefield, J. (2013) Glasgow wins 'smart city' government cash [online]. *BBC News*, 25 January. Available at: www.bbc.co.uk/news/technology-21180007 [Accessed 14 May 2013].

11
CONCLUSION

Colin McFarlane, Simon Marvin and Andrés Luque-Ayala

Smart urbanism is a loosely connected set of confluences between data, digital technologies and urban sites and processes. While we chose a sub-title for this volume that demanded a decisive normative response – utopian vision or false dawn – what the chapters in this book do instead is to ask a more modest question: what does the notion of 'smart' do? What we see is less a decisive shift in urban governance, economy, social life or environmental management, more a set of quite specific interventions in these different arenas which are best characterised as limited, often uncertain and connected to existing place-based or organisation-based logics, ideologies and debates. Part of the challenge in 'seeing' actually existing smart urbanism is the very discursive operation of smart itself, but all of the contributions in this book implore us, as a starting point, to look past the boosterism – whether that boosterism is driven by the state, corporations, civil society organisations, activists or residents – and instead critically examine why and how smart urban discourses and practice emerge and what they do and don't do. In this sense, the starting point is to open the 'black box' of smart urbanism and ask: what's really going on here and what might it amount to for urban politics, economy, environment and everyday life?

Dealing with the notion of smart is part of the challenge, then. 'Smart' is a concept that comes charged with positive and aspirational connotations. It is a radically networked concept, and pulls across a range of different discourses – economic growth, optimisation, sustainability, efficiency, better service provision, greater and more transparent citizen access, security and so on. It appears, then, as a useful and seductive concept. Who does not want to be seen as being smart? The use of smart as a concept normalises a set of aspirations and an aspirational vision of the future, even if what that future is exactly, and how it might be attained, remains somewhat elusive.

In practice 'smart' refers to a particular form of information flow: dense information usage flowing in multiple directions and offering new possibilities for recombination. Smart is not only about the discourses, but also the techniques. And some of the techniques we are dealing with here are specifically techniques around information flows (the ICT world). For example, a person is 'smart' because he or she is able to receive information in particular ways (e.g. analytically, quickly, verbally as well as emotionally), recombine it and act upon it. The feedback loop created (as he or she transforms 'something') makes this person appear 'smart'. A 'smart grid' is no other thing than an urban infrastructure network with the ability to share information in two or more ways (in contrast to traditional urban networks, where information flows only in one direction). This is achieved through the retrofit of the grid with new bidirectional devices such as smart meters, digital hubs, etc.

As we think critically about smart urbanism, it is vital, then, to consider how and with what consequences information flows within this urban configuration, and how and why information flows are being characterised as they are (Luque-Ayala and Marvin, 2014). In part, this is about unpacking urban flows and digital forms of urbanism as they exist in practice, and resisting the risk of research itself becoming incorporated within the aspirational discourses around smart. But it also means critically interrogating how smart is characterised, and here there are key theoretical, governance and methodological challenges that we think are important for urban research, and which emerge from this volume.

In closing the book, then, we wish to conclude with three reflections on researching smart – the first relates to the theoretical stakes, the second to questions of urban governance and the third focuses on the methodological challenges of researching smart urbanism.

First, theory. The question of how we might theorise smart urbanism is a searching and vital one. Clearly, it would be very wrong to attempt to legislate for which theoretical questions and approaches are important and which are not here, but we do want to argue that particular challenges stand out for theory. One important stake in the theoretical debate is to do with the *nature and interactions of technology and economy*. Here, the relation of *context* and *causality* is vital. How might we place in context the technological and economic drivers of smart urbanism? What sort of contexts? With what key drivers? This volume offers many useful answers and guidelines here. If there is a technological determinism at work in many smart city initiatives (Calzada and Cobo, 2015), there is also often too an accompanying prominence both for large multinational corporations and for a certain faith in economic efficiency as a vehicle to more effective cities (Luque *et al.*, 2014). For Hollands (Chapter 10), the technological and economic confluences need to be placed in a context of entrepreneurial governance, while for Kitchin *et al.* (Chapter 2) economic drivers (e.g. tax incentives, the marketing role of corporations and consultants) and technological dimensions (e.g. sensors and scanners, networks and dashboards) need to be understood more contingently as part of a multifaceted 'data assemblage'. Here, context is less predetermined, and causality is potentially more opaque. For example, data standards, ingrained habits, ideological leanings or local political debates and priorities might be as important as, or more important than, entrepreneurial governance. So the question of how to identify context and causality, and in particular the agency and roles attributed to the technological and economic, are and will remain important stakes in theoretical debate on smart urbanism.

Another important stake in the theoretical debate is to do with *emergence*. The smart city, as a paraphernalia of digitalised infrastructures, opens out a kind of urban presentism. The city is always – at least within the promise of smart urbanism – at hand, literally just a touch of a button away. As Nigel Thrift (2014a, 2014b) has suggested, the confluence of big data and smart technologies surfaces a great urban 'meanwhile', as streams of data about urban-life-going-on-now, or which has-just-occurred, or which is-seeming-to-appear, become more and more available to more and more people in particular kinds of visual formats (McNeill, Chapter 3). Whether this urban meanwhile takes the form of tracking sanitation challenges (Odendaal, Chapter 5), modulating energy levels (Klauser and Söderström, Chapter 7; Powells *et al.*, Chapter 8), synchronising processes and infrastructures (Gabrys, Chapter 6) or simply alerting urbanites to changes in traffic or new spots for urban nightlife (Kitchin *et al.*, Chapter 2), it challenges us as researchers to conceptualise the ways in which urban perception, experience, politics, ethics and life may or may not be altered by the presentism of an emergent urbanism. Whether this adds up to new (visual) ontologies of the city based on the visual (McNeill, Chapter 3) or the anticipatory (Klauser and Söderström, Chapter 7) or the experimental (Powells *et al.*, Chapter 8; Calvillo *et al.*, Chapter 9), or simply new ways of doing old processes, or new registers of urban epistemology, is a fascinating debate attached to this concern.

Taken together the relations between *emergence*, *causality* and *context* raise important questions about how to characterise 'smart'. Does smart urbanism, a convenient shorthand to be sure, 'add up' to a generic process as such? What is often striking about smart urban initiatives run by city councils is the level of contingency and uncertainty sometimes found around what smart is and how it might evolve. Our experience, and we see it across the chapters in this volume, is that there is less a sense of clear vision around smart – despite what the glossy websites and videos of the urban future might suggest – and more a sense of bumbling through; a sense that 'smart matters' but without any real clarity about why and how, or any certainty about how and why to get different groups working together, from government departments to civil society groups, private companies and ordinary residents. If there is a generic sense of smart here it often lies in a sense of following a kind of 'smart script', from developing control rooms (often with Rio de Janeiro in mind) to sponsoring hackathons and simply getting data 'out there' in the public domain. Here, smart emerges not so much as a linear logic but more haphazardly as a domain that cities must be seen to be involved in; and of course, there is often opportunity for raising varied resources in this area. Cities might move from funder to funder, demonstration project to demonstration project, and the promise of the smart urban revolution is that it is always almost there – emergent, yes, but rarely quite arrived. This loose, contingent, broadly shared script of smart is shared from Cape Town to Glasgow.

The second issue is that we need to ask what kinds of urban governance might be produced by smart urbanism. Does smart urbanism represent a new way of governing the city, or simply a new iteration of existing forms of governance? Or are the forms of power and governing at work in the category we are currently choosing to label 'smart' simply too manifold and diverging to be distilled into particular logics or rationales of governing?

Across this volume, smart urbanism is connected to a wide range of urban governance discourses and practices, identifying in the process important avenues for critical urban research here. Kitchin *et al.* (Chapter 2) identify, amongst other logics, a *technocratic* vision of urban governance that reduces urban problems to technological and data-driven calculations that require central command-and-control procedures. Oftentimes, this technocratic rendering of urban governance, which echoes claims that governance effectively marshals a kind of post-political condition (Swyngedouw, 2009; MacLeod, 2013), is accompanied by a particular role for the visual. As McNeill (Chapter 3) argues, part of the promise of a lot of smart city discourse and practice is being able to 'see' the city more completely and in real time, and in ways that might enhance capacities for supervision and foresight.

Hollands (Chapter 10) examines the pervasive presence in mainstream smart urban initiatives of a kind of *entrepreneurial* vision of urban governance, where becoming 'smart' is to relinquish more and more control to business-led development and tech-savvy creative citizens. This is a case made powerfully by Datta (Chapter 4), who examines the interplay between smart city initiatives, entrepreneurialism and neoliberalism in India. Powells *et al.* (Chapter 8), too, locate entrepreneurial models of urban governance in smart grid initiatives. Or, to take another example, Klauser and Söderström (Chapter 7) focus on smart as an attempt to govern the city through code as a broadly cast *security problematic*, where the aim is to securitise, regulate and manipulate particular kinds of flow such as those of energy in relation to shifting patterns of energy production, movement and use. There are implications too here for the kinds of citizenship these modes of governance might entail. For example, Gabrys (Chapter 6) argues that what is governed in the intersections of smart and sustainable urbanism is less the citizen as a subject and more citizenship – or 'environmental citizenship' – through data distribution, feedback and monitoring (a form of biopolitics 2.0).

Such imaginaries of urban governance – technocratic, entrepreneurial, security-orientated – may well be commonplace to smart city discourses and practises, but this volume also suggests that (a) they are by no means exhaustive of the forms of governing that smart urbanism may entail and (b) far from being totalising logics, they are more often only partially worked through or operationalised. So, Hollands for example does not only connect smart city initiatives to forms of urban neoliberalisation, but also traces important alternative forms of participatory smart urbanism that point to a more democratic and community-orientated kind of technology (Chapter 10). Equally, Kitchin *et al.* show that while key smart governance technologies such as indicators, benchmarking and dashboard initiatives too quickly construct questionable notions of 'truth' despite their technical and methodological limitations and exclusions, they nonetheless provide data that can be used by any number of groups in ways that are socially emancipatory and empowering. As they argue, 'data assemblages' are made through particular contexts, and can open out unexpected insights and possibilities (Chapter 2).

Several of the contributors are keen to make this claim: that in practice smart urbanism is not a closed but potentially an open governance agenda, albeit in the context of often powerful vested interests. Moreover, smart urbanism in practice is quite often only partially embedded in governance practices. The extensive control room systems we see in cities like Rio de Janeiro are exceptional (Luque-Ayala and Marvin, forthcoming); more often, smart city programmes are highly situated, specific and disparate across different groups and agendas. A city council like Glasgow – a test bed for UK smart cities – may well have an impressive control room, albeit one that is far more limited than in the Rio case, but many of the activities of the smart city initiative are not generalised across the council itself let alone other groups in the city, whilst at the same time other non-governmental groups are able to put digital technologies to uses that far exceed the city's formal smart city initiative. Odendaal (Chapter 5) shows, in relation to Cape Town, that smart city initiatives are fragmented, only partially developed and with quite different agendas attached to them, some of which are potentially progressive for ordinary citizens (such as the Social Justice Coalition's smart phones project that monitors and reports on urban sanitation conditions).

While critical urban theory and research needs to expose, critique and develop alternatives to neoliberal, technocratic, positivist and surveillance-orientated forms of smart urbanism, it is just as important for critique not to imply that these versions of the smart city are more developed, pervasive and in control of urban governance than they actually are. Indeed, given the often partial and fragmented nature of smart urbanism discourses and practices, it may be too early to talk of 'smart urban governance' – such as through a single coherent 'smartmentality' (Vanolo, 2014). Certainly, it is too early to attribute the status of a 'new way of governing' to smart city initiatives. Instead, the urban governance question remains rather more open, there to be populated by a range of groups and voices even if in a context of uneven power relations.

Its possible, then, that academic researchers have more potential to populate the vision and practice of smart urbanism than we may assume. This takes us to our third concern here: methodology. The chapters in this volume present a range of different ways in which we might research the smart city. Four key ways into researching the smart city emerge in the book, and many of the authors adopt several of these at once. Smart is researched through a focus on: (i) *discourse and imagery* at different levels, including policy, city council, corporate organisation or activist groups (McNeill, Chapter 3; Datta, Chapter 4; Odendaal, Chapter 5; Calvillo et al., Chapter 9; Hollands, Chapter 10); (ii) *description and evaluation of flagship case studies*, from Songdo (Calvillo et al., Chapter 9) and IBM (McNeill, Chapter 3) to the case of Dholera as part of India's wider smart city effort (Datta, Chapter 4) or the MIT-Cisco Connected Urban Development initiative (Gabrys, Chapter 6);

(iii) the *identification of a key trend* across different sites, from dashboards in Dublin and London (Kitchin *et al.*, Chapter 2) to a focus on particular kinds of active and anticipatory citizenship in smart sustainability projects (Gabrys, Chapter 6; Klauser and Söderström, Chapter 7) or the seductive work of the very label 'smart city' (Hollands, Chapter 10); and (iv) *immersion in the life of complex smart city projects*, where the researcher is sometimes actively involved in the constitution of what smart is in practice (Kitchin *et al.*, Chapter 2; Powells *et al.*, Chapter 8).

Clearly, all of these different methodological routes into smart – *discourse*, *case study*, *key trend* and *immersion* – are important and offer different kinds of detail and generalisation. What we find particularly exciting about these routes is that they all offer the opportunity not just to critically evaluate smart urbanism, but to populate the future of what smart urbanism is and might be. Particularly important here is the example of *immersion* with in actually existing projects, where there is a potential to bring a more socially and ecologically aware smart into being, but there is an important role, too, for work that critically engages with the discursive framing of smart urbanism.

A second set of methodological challenges relates to the status of the category of 'urban' in 'smart urbanism'. As several of the chapters in this volume have noted, critical research needs to ask what kind of urbanism is at work in these discourses and projects. Does the discourse of smart in India, as Datta critically explores in Chapter 4, begin with global abstractions of the value of smart cities, or with the needs and realities of urban life in India? Odendaal in Chapter 5 traces efforts to connect smart urbanism to the realities of life in Cape Town, just as Kitchin *et al.* in Chapter 2 cautiously approach what dashboards may and may not offer different constituencies in Dublin. Methodologically, the challenge here is to ask whether and how the specificity of urban life features as constitutive of smart, and this entails a research engagement not just with smart discourses and projects but with the putative 'urban' that those discourses and projects target. The methodological challenge here is not to engage the urban simply as a space, but as a logic, imaginary and form – whether as a 'test bed' (Calvillo *et al.*, Chapter 9) or in its relation to a transurban smart grid (Powells *et al.*, Chapter 8) or through a series of forms of sensing (Gabrys, Chapter 6; Klauser and Söderström, Chapter 7) or as a set of visual codes (McNeill, Chapter 3) or as a potential site of alternative urbanisms (Hollands, Chapter 10). The different ways in which the urban surfaces in smart initiatives is, then, not only a theoretical question, but a methodological challenge of working through how we might position both the urban that is framed by smart and the urban that is marginalised or actively excluded from it.

References

Calzada, I. and Cobo, C. (2015) Unplugging: deconstructing the smart city. *Journal of Urban Technology* 22 (1): 23–43.

Luque-Ayala, A., McFarlane, C. and Marvin, S. (2014) Smart urbanism: cities, grids and alternatives. In M. Hodson and S. Marvin (eds) *After Sustainable Cities?* New York: Routledge, pp. 74–89.

Luque-Ayala, A. and Marvin, S. (forthcoming) The maintenance of urban circulation: an operational logic of infrastructural control. *Environment and Planning D: Society and Space.*

Luque-Ayala, A. and Marvin, S. (2015). Developing a critical understanding of smart urbanism. *Urban Studies* 52(12): 2105–2116.

MacLeod, G. (2013) New urbanism/smart growth in the Scottish Highlands: mobile policies and post-politics in local development planning. *Urban Studies* 50 (11): 2196–2221.

Swyngedouw, E. (2009) The antinomies of the postpolitical city: in search of a democratic politics of environmental production. *International Journal of Urban and Regional Research* 33 (3): 601–620.

Thrift, N. (2014a) The promise of urban informatics: some speculations. *Environment and Planning A* 46 (6): 1263–1266.

Thrift, N. (2014b) The 'sentient' city and what it may portend. *Big Data & Society* 1 (1): 1–21.

Vanolo, A. (2014) Smartmentality: the smart city as disciplinary strategy. *Urban Studies* 51 (5): 883–898.

INDEX

ABI Research 171–2
actor network theory 43
Adey, P. 46, 116
advertising 174
Agenda 20, 23
Ahmedabad 57
air-source heat pump (ASHP) technology 133–4
algorithms and algorithmic processes 116, 160–1, 164–5
Ambani City 54
Amin, Ash 163
Amsterdam 171, 178
Anderson, B. 46
Anttiroiko, A.-V. 172, 174, 180
Appadurai, Arjun 158
Arup (company) 47
'assemblages' of data 20–2, 125–30, 141–3, 186–9; apparatus and elements of 20
Austin (Texas) and Austin Energy 131–2, 135–7, 140–1
Australia 137
automation 102, 115, 151–4

Bacon, Francis 161
Baltimore 25
Bangalore 54, 74
Bangalore–Mumbai corridor 61
Bangalore–Mysore corridor 57, 62
Barcelona 55, 171
Bayesian inference 159–60
benchmarking of cities 23–9
'big data' 5–6, 19, 34–5, 41, 46–9, 146, 156, 160

biopolitics 90–3, 103–4
BKW (company) 109–10, 114, 118, 120
Bladerunner (film) 169–70
Blair, Tony 49
Block, T. 25
Blok, A. 125
boosterism 185
Borpuzari, P. 55–6
Boston 55
Brickstarter 177
bridges.org (NGO) 71–2
British Gas 133–4
Brown, L. 129
Budd, L. 116
Bulkeley, Harriet viii; *co-author of Chapter 8*
Bunnell, T. 65
business involvement in smart cities 172–6, 179–81

Calvillo, Nerea ix, 10, 187–90; *co-author of Chapter 9*
Cape Town 9, 71–84, 189–90
Carter, S. 44
Cary, Jonathan 35
Casalegno, Federico 89, 93–6, 100, 102
Central Electricity Generating Board 128
Centre of Regional Science 170
Chakrabarty, D. 54–5, 65
Chicago 7
Choi, Gui-Nam 146
Choplin, A. 54
Cisco (company) 9–10, 37, 54, 59, 71, 73, 89, 100, 104, 148–50, 153–4, 162, 169, 172–3

citizenship 103–5, 188–90
City 2.0 92
climate change 74–5, 83, 92–3
code, governing by means of 108–10, 115–21, 188; related to normalisation 117–19; related to space 119–21
collective behaviour of a city 154
Comaroff, J. 54
community involvement 139, 175–9, 188
Compstat technology 40
Connected Sustainable Cities (CSC) project 89–94, 97–8, 102–4
Connected Urban Development initiative 189
Cook, Justin 47
Copenhagen 171
Crampton, J.W. 112
Crary, J. 38, 40
Cukier, K. 45
Curitiba 94, 101–3
customer-led network revolution (CLNR) 130, 133–4, 139
cybernetics 45

dashboards, digital 5–6, 24–30, 38–41, 44, 189–90
Data Control Company 132–3
data mining 151, 154
data problems 27
Datta, Ayona ix, 188–90; *author of Chapter 4*
de Lange, M. 176, 178
Deleuze, G. 105, 138
Delhi–Mumbai industrial corridor (DMIC) 56, 73–4
democratic decision-making 52, 64, 175
de Waal, M. 176, 178
Dholera 10, 53–66, 189
digital divide 73, 79, 82
'discipline' concept (Foucault) 110–13
'dispositif' concept 22, 89, 92, 97, 104, 126
distribution network operators (DNOs) 129, 134, 139
Dublin Dashboard 21–2, 28–30, 189–90
Dubuque 37
Durban 71

ecological fallacy 27
Edison, Con 45
Eizenberg, E. 179
Elden, Stuart 112, 161
electricity meters 109, 120, 137–9
electricity supply: balancing of production and consumption 10, 114–21; *see also* smart grids in energy networks

'energy dilemma' 126
energy networks 10, 127; *see also* smart grids
entrepreneurial urbanisation 53, 56, 60, 62, 65–6
entrepreneurialism 56–9, 174, 188
environmentalism 91–2, 104
environmentality 9, 89–93, 102–5
Equilibrium (film) 169–70
Ericson, R. 120
Eurocities 172
European Smart Cities Initiative 171
European Union (EU) 127, 129, 133

Face your World project 178
Fairclough, Norman 49
596 Acres project (Brooklyn) 178–9
Flanders 26
flexible consumption of energy, geographies of 142
Flexlast project 109–21
Foucault, Michel 9–10, 22, 89–93, 102–5, 109–16, 119–21, 126, 161
Franck, A. 54
Fremantle 37
French, S. 116
Fujitsu (company) 169, 173
Future Cape Town (think-tank) 77

Gabrys, Jennifer ix, 5, 9, 187–90; *author of Chapter 6*
Gale, Stan 173
Gale International (company) 148, 172
Gates *Access to Learning* Award 79
GIFT 54–5, 66
Giuliani, Rudy 40, 49
Glasgow 171, 189
global positioning system (GPS) technology 94
Goldman, M. 57, 61–4
Google 104
governance, corporatisation of 18
governmentality: *automated* and *anticipatory* 115–16; Foucault's concept of 110
Graham, S. 44, 129, 140, 175
greenhouse gas emissions 171
Guangzhou Knowledge City 171
Guattari, F. 138
Guayaquil (Ecuador) 162
Gujarat 54–7, 60–5
Guy, S. 128

Hacking, I. 154–5
Haggerty, K. 120
Halcrow (consultancy firm) 54, 57, 59, 61

Halpern, Orit ix; *co-author of Chapter 9*
Hamburg 94, 98
Haollenphai 66
Haraway, Donna 163
Harvey, David 137, 174
Harvey, P. 47–8
Helsinki 171
Hermant, E. 43–4
heterotopias 147
Hewlett Packard (HP) 104
Hezri, A.A. 28
Hill, Dan 172
HITEC CIty 54
Hodson, M. 133
Hollands, Robert G. ix, 10, 60, 130, 148, 186–90; *author of Chapter 10*
Holston, J. 65, 67
Hoornweg, D. 179
Hubert, C. 138
Hudson, R. 23, 174, 177
100 Smart Cities programme 52–7, 60, 65
Hurricane Katrina 92
Hyderabad 74

IBM (company) 6, 8, 34, 40–9, 71–4, 100, 110, 148, 169, 173, 189
Immersive Labs 174
Incheon Free Economic Zone (IFEZ) 145–50, 156
India 9, 16, 52–67, 190
indicators, urban 22–30
informal dwellings 75
information and communication technologies (ICT) 4–6, 16–17, 55, 72–9, 83–4, 98, 103, 127, 129, 137, 168–72, 179–80
infrastructure, supervision and inspection of 44–6
instrumental rationality 29
Intelligent Community Forum 171
interdisciplinarity in urbanism 2
International Pirate Party 176
internationally comparative analysis of smart urbanism 2–3
Internet access 134
interrelation, governing by means of 114–15
iSMART project 109–21

Jameen Adhikar Andolan Gujarat (JAAG) 63–6
James Lang LaSalle 24
Jessop, B. 52, 56
Jongwon Kim 145, 148

Kalpasar dam 57, 62
Kant, Amirabh 59
Kargon, R. 53–4
Kasarda, J. 147
key performance indicators (KPIs) 40–1
Khayalitcha 75, 79–83
Kim, Tony 146
Kirby, T. 171
Kitchin, Rob ix, 8, 34, 41, 186–90; *co-author of Chapter 2*
Klauser, Francisco R. ix, 9–10, 187–90; *co-author of Chapter 7*
Knight, Frank 15, 69
Korea Telecom 148, 174–5
Kyung-Sik Chae 145–6

land acquisition 53, 57–66
LandProp 173
Latour, B. 39, 43–4
Lauriault, P. ix–x; *co-author of Chapter 2*
Lavasa 54–5, 66
Law, John 41–2
LeCavalier, Jesse xi; *co-author of Chapter 9*
Le Corbusier 162
Levien, M. 62
libraries 79, 84
Lindsay, G. 147
Living PlanIT 148
London 128
London Dashboard 25, 189–90
Low Impact Living Affordable Community (LILAC) 177–8
Luque-Ayala, Andrés ii, viii, 52, 54; *co-editor and co-author of Chapters 1 and 11*

McAdams, M.A. 179
McArdle, Gavin xi; *co-author of Chapter 2*
McFarlane, Colin ii, viii, 54, 125; *co-editor and co-author of Chapters 1 and 11*
McGuirk, J. 177
Mackenzie, A. 98
MacKenzie, Donald 157
McLean, Anthony xi; *co-author of Chapter 8*
McNeill, Donald xi, 8, 187–90; *author of Chapter 3*
Madrid 94–5
Malaysia 65
management-at-a-distance 108–9
management-consulting services 148
managerialism 23–6
Manchester 47, 171
Manesar-Dhow 66
marketing language 73–4, 83
marketisation of energy systems 132

Marvin, Simon ii, viii, 52, 54, 129, 133, 175; *co-editor and co-author of Chapters 1 and 11*
Marx, Karl 170
Masdar 164, 168, 173
Massachusetts Institute of Technology (MIT) 9, 89
Massumi, B. 92
Mattern, S. 18
Mayer-Schönberger, V. 45
Mbembe, A. 163
Meyerson, Bernie 47
Migros (company) 109, 114
Minority Report (film) 169–70, 174
Minton, Anna 173
Mitchell, H. 82
Mitchell, W. 89, 93–6, 100, 102, 154
modelling 47, 116
Modi, Narendra 52, 60, 73
Mokos, J.T. 6
Mollela, A. 53–4
Monahan, T. 6
monitoring processes 96, 100, 103–4, 151
Morozov, E. 18
multinational corporations 18, 71, 186
Mumbai 54
Mumbai–Bangalore corridor 61
Murray, Tania 39
Mysore–Bangalore corridor 57, 62

Narmada dam 57, 62
NEC (company) 174
network structures 120
new public management 41
New York 40, 45, 49, 178–9
normativity and normalisation 28–9, 111–12, 117–21; *see also* performative normativity

objectivity (supposed) of indicator, dashboard and benchmarking initiatives 26–7
Odendaal, Nancy xi, 9, 187–90; *author of Chapter 5*
Olympic Games 74
Ong, A. 54
open data platforms 7
optimisation mechanisms 10, 114–15, 119–21
Osborne, T. 34, 39
Osorio, Carlos Roberto 38
Otter, Chris 35, 39, 41, 44–8

Paes, Eduardo 35–8
Parasuraman, Krishnan 46
Paris 39, 43

'pay-as-you-go' electricity supply systems 140
Pecan Street Project (PSP) 126, 131, 135–6, 139–40
Peck, J. 54, 61
performative normativity 154–6
Pietsch, Wolfgang xi; *co-author of Chapter 9*
Pike Research 171–2
PlanITValley 173
political leadership 49
politics of urban data 5–6, 19–22, 26–30
pollution 103
Portland 47
Portugal 148, 173
Powells, Gareth xi, 11, 187–90; *co-author of Chapter 8*
predictive techniques 19, 47
privacy laws 153
probabilistic thinking 159
programmed environments 97
'prosumers' 10, 120, 136–7, 140–2
Provoost, M. 176
public land, use of 174, 178
public–private partnerships 132–7, 141–2, 149–50, 175
public services, financing of 135–6, 175

Rajarhat 66
ranking of cities 27, 170
realist epistemology 29
reflexive questions on smart urbanism 4
regulatory regimes 121
renewable energy 127
Rio de Janeiro 6–8, 23, 35–8, 47, 55, 74, 83, 187, 189
risk as distinct from uncertainty 158–60
Rose, N. 34, 39
Roy, Arundhati 54, 56

San Francisco 89, 98
SAP (company) 78, 83
Sassen, S. 61
score-carding 23
'security' concept (Foucault) 110–18, 121
self-regulating citizen activities 103
Sennett, Richard 168, 179
sensor technologies 5, 90, 93–104, 151–4, 157, 190
Seth, Sunil 74
'shadow data' 154
Shendra-Bidkin 66
Sheppard, E. 54
Siemens (company) 37, 77, 83
simulation 47, 116
Singapore 37, 47, 73–4, 171, 175

Singer, N. 35–6, 73
Smart City Council 56
smart grids in energy networks 125–43, 175, 190
smart poles 151–2
smart urbanism: contrasting views of 170–1; critiques of 3–4, 8, 17–22, 29, 168, 172–5, 187; drivers of 186; existing and emerging debate on 4; and governance 187–9; research on 2–8, 189; theorisation of 186; visions of 16
smartphones 72
Social Justice Coalition (SJC) 76, 80–4, 179
social sorting 19
socio-technical apparatus and processes 6, 78, 126, 133, 142, 176
Söderström, Ola xi, 9–10, 45–6, 187–90; *co-author of Chapter 7*
Songdo 11, 89, 145–6, 168–74, 189
South Africa 9, 71; *see also* Cape Town
South Korea 10, 145, 162; *see also* Songdo
South Tyneside Homes (STH) 134
spatiality 112–13
state institutions, role of 126–7, 132–3
Stockholm 175
Strand East 173
strategic essentialism in indicator, dashboard and benchmarking initiatives 30
Stratford (Ontario) 171
Strengers, Y. 136
'striations' (Deleuze and Guattari) 138–9
Sum, N. 52
Sunderland 171, 175
surveillance 19
sustainability 89–90, 93, 96–100, 103–4, 149, 171–2
Switzerland 10, 108–10

technocratic visions of urban governance 17–18, 165, 188
technological determinism 186
Terminator films 169–70

Terranova, Tiziana 163
territory, new concept of 161
test-bed urbanism 154–65; definition of 147; about population and territories 161–2
Thatcher, Margaret 26
Thessaloniki 171
Thrift, Nigel 44, 113, 116, 187
Toshiba (company) 96, 104
Townsend, A. 47, 52
transition to smart urbanism 2

ubiquitous cities ('u-cities') 170
ubiquitous computing 89–90, 93, 96, 100, 161
uncertainty, statistical calculation of 158–60
United Nations Conference on Environment and Development (UNCED) 23
urban operating systems 6–7
user-generated content 92
utopian visions 160, 163

Valentine, S. 79
Vanasi 55
Van Assche, J. 25
Vanolo, A. 4–5, 53
Venus Project 176
Verbong, G. 130
visualisation 35–41, 49; as introspection 39
vulnerability of digital systems 18–19

Watson, V. 54, 61, 65, 73
World Cup football 74
World Forum on Smart Cities (1997) 148

YIMBY projects 177

Zeitgeist Movement 176